CARL O. SAUER

A Tribute

LIHC

CARL O. SAUER
A Tribute

edited by

Martin S. Kenzer

Published for the
Association of Pacific Coast Geographers
by the
Oregon State University Press
Corvallis, Oregon

The paper in this book meets the guidelines for permanence and durability of the Committee on Production Guidelines for Book Longevity of the Council on Library Resources.

Library of Congress Cataloging in Publication Data
Carl O. Sauer—a tribute
   "The published writings of Carl Sauer": p.
   Includes bibliographical references and index.
   1. Sauer, Carl Ortwin, 1889-1975—Congresses.
2. Geography—Congresses. 3. Geographers—United States—Biography—Congresses.
I. Sauer, Carl Ortwin, 1889-1975. II. Kenzer, Martin S., 1950-
G69. S29C37   1986   910' .924   86-19184
ISBN 0-87071-248-9 (alk. paper)

Contents

Part Six
A Summary

Carl Sauer:
Explorer of the Far Sides of Frontiers
Alvin W. Urquhart
217

Foreword

This is a book about a remarkable man whose contribution to the history of geography in North America remains the subject of ongoing appreciation and assessment.

When Carl O. Sauer was born in Warrenton, a quiet German community in Missouri, in 1889, William Morris Davis was thirty-nine years of age. North and South Dakota, Montana, and Washington were admitted as states into the Union, and Frederick Jackson Turner was soon to announce his frontier thesis. Geology had been brought into the service of the government, surveys had been completed, mapping programs were advancing, and mineral wealth was being appraised. Academic studies in physiography and ethnology were emerging in the universities as geology evolved into a study of its most recent stage. During the eighties Davis had gone west to the dry lands, where every cowboy was a geologist, there to develop the concept of the cycle of erosion. Physiography now had an organizing concept, and discipline was possible. Davisian physiography, institutionalized in The Committee of Ten Report, became the foundation for a geography which would shortly be humanized in causal posture. Varieties of determinism, influence, adjustment, and the geographic factor were soon to be the subject of scrutiny. This was the intellectual milieu which prevailed during the earliest years of Carl Sauer.

There in Warrenton, Missouri, he grew up close to nature, with the habits of rigorous education present both at home and at school. At the age of nine he began three years of schooling in Calw at the eastern foot of the Schwarzwald, Germany, then returned to graduate from Central Wesleyan College (Warrenton). Thence he went to Northwestern University to study geology but after one year opted to study geography at the University of Chicago, which boasted the only Department of Geography in North America then offering a doctorate. (Harvard and Yale Universities had awarded doctorates in geographical subjects, but these were bestowed by the departments of geology.) Sauer respected the learning and questioning mind of Rollin D. Salisbury, but for the remainder of the Chicago geographers he had less enthusiasm. Pedestrian minds, teaching without inspiration and infatuated with the causative notion of environmental dominance, discouraged Sauer. He withdrew from his program of study to work for some months with the Rand McNally Company. While removed "from all the little Salisbury's around the building" he began to read German geographical literature:

> My dissatisfaction with the environmentalist tenet came mainly from listening to Miss Semple and J. Paul Goode, both delightful persons, and hearing Barrows distinguish between geographic and non-geographic factors. That wasn't what I had come for to Geography. In the years I worked in The Loop I read German geographers evenings who were doing what I wanted (*letter from C. O. Sauer to William W. Speth, March 3, 1972*).

The presence of a German component in North American geography was already noticeable. Americans who had studied geography in Germany prior to 1892 (the year of The Committee of Ten) included Cleveland Abbe Sr., Richard T. Ely, William H. Hobbs, Lindley M. Keasbey, Charles T. McFarlane, Charles A. McMurry, Francis W. Parker, Rollin D. Salisbury, Edward Van D. Robinson, and Edward L. Stevenson. Other American geographers who studied in Germany after 1892 included Ellen C. Semple (under Friedrich Ratzel, 1892-93, and 1895), Charles T. McFarlane (Albrecht Penck, 1898), J. Russell Smith (Friedrich Ratzel, 1901-02), Wellington D. Jones (Alfred Hettner, 1913), Eugene van Cleef (Joseph Partsch, 1913-14), and Samuel Van Valkenburg (a later participant in American geography who studied with Albrecht Penck, 1915-16). Robert DeCourcy Ward had spent some of his youth in Germany, while his father was with the diplomatic service. Carl Sauer had wanted to study at Heidelberg University under Alfred Hettner but circumstances had prevented this *(Gottfried Pfeifer to Mrs. W. Hess, January 9, 1956)*. And several American geographers had corresponded with German geographers, particularly Jean L. R. Agassiz, William M. Davis, Daniel C. Gilman, Arnold H. Guyot, John W. Powell, Ellen C. Semple, Nathaniel S. Shaler, Robert DeC. Ward, and George M. Wheeler. German and American geographers also met on the occasions of the International Geographical Congresses. There was legitimacy to Hettner's claim that geography was a German science. Undoubtedly there was a strong German component in American geography, the study of which is still undone.

There was also a large indigenous geographic component inspired by the wilderness-conquest experience and the work of the Geological Surveys. Much of this work had been synthesized and systematized by Davis who provided order and perhaps the beginnings of geographic discipline with his concept of the cycle of erosion. Man was then added to the physiographic base; ontography had been added to physiography. A causative mode was produced, and a creative period of American geography ensued. Departments of geography were formed, geographical societies increased in size and number. The first department of geography in North America to offer a doctoral degree was that of the University of Chicago in 1903, the Association of American Geographers was founded in 1904, and a peripatetic International Geographical Congress was held in Washington D.C. that same year. Ten years later the National Council of Geography Teachers was formed (largely at the urging of George J. Miller), the National Geographic Society had celebrated its twenty-fifth anniversary and already possessed membership nearing one million, and the American Geographical Society brought Isaiah Bowman to its leadership (1915). Geography was under spreading sail. Even World War I was to provide geography with numerous opportunities, e.g., by bringing The Inquiry to the rooms of the American Geographical Society and ensuring the participation of a number of geographers in the negotiations leading to the Paris Peace Conference. Of this activity Sauer could write, "It was a springtime, the only good one we've had in this country" *(C. O. Sauer to L. S. Wilson, April 6, 1948)*. Intellectual curiosity was at a high point, there were a lot of good minds simultaneously at work in the field, and America was expansive and confident in its mood. Geologists, geomorphologists, and geographers interacted and benefited from knowledge and sense of problem developed by each other. Geographers were schooled in matters of the fundament and human occupance. There was much to be done and geographers set about doing it. Davis at Harvard was responsible for molding the first generation of professional geographers. This primal generation included (but is not restricted to) Isaiah Bowman, Albert P. Brigham, Alfred H. Brooks, Richard E. Dodge, James W. Goldthwait, Ellsworth Huntington, Mark S. W. Jefferson, Curtis F. Marbut, Lawrence Martin, Vilhjalmur Stefansson, Walter S. Tower, and Robert DeC. Ward. These were makers of early twentieth-century American geography. They assumed university and normal school positions across the country and took with them Davisian learning.

In 1914 Sauer was employed at the Salem Normal School, Massachusetts, where he developed a friendship with Sumner Cushing, a one-time student of Davis. During that summer Sauer taught at the University of Michigan, under the direction of William H. Hobbs, who had been head of the Geology Department since the death of Israel C. Russell in 1906. In January 1916, Hobbs hired Sauer as an instructor in geography, though apparently he would have preferred to hire T. Griffith Taylor *(W. H. Hobbs to T. G. Taylor, November 26, 1935).* Sauer developed the curriculum, ordered geographical literature for the library, embarked on a publication record, developed an acquaintance with Mark S. W. Jefferson (seven miles distant in Ypsilanti), and won for himself a reputation as "outstanding." "Sauer was an engaging lecturer with most remarkable mustache and eyebrows, a villainous pipe, and a pocket full of kitchen matches. Even without the smoke screen, Sauer would have been outstanding."[1] He published a number of articles in the journals, and read voraciously in the literature. Sauer was not an enthusiast of the geographical cycle or the causal posture, but this was not made very apparent during his stay of nearly eight years at Ann Arbor. Possibly this was because the departments at Chicago and Michigan were largely devoid of Davis's students. Neither Salisbury nor Hobbs was "Davisian," and both had studied in Germany. Possibly this had influenced Sauer's thinking (away from Davis and toward that of Salisbury).

Sauer was elected to the Association of American Geographers in 1920 (proposed by W. H. Hobbs and R. D. Salisbury), and at once held positions on several association committees: "Delimitation of Geographic Regions," "Geographic Illustrations" (chairman), "Geographic Education" (chairman), "Advisory Committee on Agriculture Geography." He selected a site for the Geography Department's field camp at Mill Springs, Kentucky, and helped to initiate a survey of the cutover lands of northern Michigan. Both enterprises demonstrated thought and insight and both were successes. Hobbs had encouraged the development of a geographical coterie within a geology department—in 1936 Hobbs was elected President of the Association of American Geographers—and by 1923 had provided Sauer with a professorship and a department of geography.

In that year Sauer chose to accept a position at the University of California at Berkeley; it was a position to which Derwent Whittlesey had been invited but which he had declined (his close friend Miss Semple had urged him to go east, which she felt was more "civilized"). Sauer later wrote:

> . . . the summer obligations were one of the reasons that made me choose California. Between the University summer camp and the Michigan Land Economic Survey I saw myself tied down for an indefinite time into the future and I wanted to get out on my own and roam new areas. . . . I don't recall that I ever taught a summer session since I came out here *(C. O. Sauer to Wellington D. Jones, November 9, 1951).*

Sauer went to Berkeley taking with him a very talented student, John B. Leighly. At Berkeley, Sauer developed a department in his own image. Faculty were selected with this in mind (no distortions were endured pursuant to any political mode), and students selected a genre of geography available nowhere else in North America. It was characterized by the long view of human occupance of the earth's surface, inscriptions, centers of innovation, and dispersal of their cause. It was an enterprise particularly associated with fieldwork to which Sauer brought unusually powerful analytic command devoid of the impressionistic hypotheses of some previous workers. It was also characterized by a determined belief that a human geography could not be separated from the physiography of its hearth. Sauer joined with Salisbury in appreciating the role of time as an indispensable attribute of physiographic reality, and in so doing distanced himself from what he considered to be the theoretical properties of the Davisian model of the cycle. Later Sauer was to write that "The influence of W. M. Davis froze the entire content of geomorphology a generation ago in this country, and there has been no freeing of those bonds except for a little shaking that we have done here" *(C. O. Sauer to Erich W. Zimmerman, May 19, 1937).* And again, "W. M. Davis has done a tremendous job, but to his disciples he has written the decalogue, and for that reason physiography is a practicing profession, not a scientific inquiry, open minded as to revolutionary hypotheses" *(C. O. Sauer to Augustus C. Trowbridge, December 18, 1935).* Too, Sauer bestowed upon culture the active role in an inversion of the determinist position which had bestowed upon fundament that same active role. This was a bold conceptual opposition to the prevailing notion of a dominant and determining fundament. In a terse expression of his belief he wrote to Wellington D. Jones, "I do think that our scientific end is in understanding how things came to be. Physical geography does seem to me to be natural history and human geography culture history" *(January 31, 1934).*

Structure to Sauer's geographic undertaking was revealed in part in "The Morphology of Landscape," a form of manifesto. Elsewhere the writer has suggested that this article replaced the Davisian physiography and causation paradigm with a new direction characterized by the sobriquet, "field and region."[2] Approximately 60 percent of Sauer's references in the "Morphology" were to German sources: the term and method both owe to Carl Ritter. Leighly has written that at this time Sauer did not know Schlüter's work, but was probably most influenced by Passarge, though he was reading Keyserling, Vaihinger, and Goethe, and Spengler when writing the "Morphology" *(John B. Leighly to Richard Hartshorne, November 6, 1975).* While Sauer later disowned his methodological writings, this essay did much to bring attention to German geography in the United States. He wrote to Alfred Hettner that it was only from a distance that he had "come to value German geography and to present certain parts of it here in the U.S." *(January 7, 1926).* And he later wrote to William W. Speth that after reading in German geography he "put it together as the Morphology of Landscape" *(March 3, 1975).* The younger geographers were attracted: Leonard S. Wilson has written "my generation of geographers was weaned on it" *(L. S. Wilson to C. O. Sauer, April 15, 1948);* and Robert E. Dickinson, referring to it as an "(obtuse, but provocative) essay," regarded it as "my springboard into geography as a 'discipline'" *(R. E. Dickinson to T. Walter Freeman, July 8, 1980).* Preston E. James has written of it as a bugle call to the young geographers, and in 1940 wrote:

> If some of the elder statesmen could provide . . . the coherent point of view which you set forth in 1925 in the "Morphology of Landscape," but now revised in the light of later criticism, nothing more important would have been done to develop the kind of coherence. . . . to guide the next 10 years of American geography *(P. E. James to C. O. Sauer, March 22, 1940).*

Oskar Schmieder, then a member of the Berkeley Department, wrote to Alfred Hettner:

> The "Morphology of Landscape" does not tell us Germans a lot of new things, was not intended to do so, but for the first time in American literature focuses on the term landscape. I do not believe that ever before anything of equal value was produced in this country (*January 1, 1926*).

But the older generation of American geographers was little moved by the essay. Dryer, in reviewing the work, challenged the methodology of going into the field without working hypotheses. "If he (the geographer) does, the result is likely to be a catalogue half rubbish, like a child's collection from a dump heap, and wholly unscientific."[3] Yet arguably it was from such thought that Sauer was to make his contribution to the literature. That contribution included essays on the objectives and methods of geography, on regional synthesis, field studies in geomorphology of the Far West (designed to test the merits of the Davisian and Penck positions), archaeogeographic studies, historical geography with special reference to the American Southwest and north Mexico, and essays on conservation and settlement. It also included especially thoughtful essays on agricultural origins and dispersals, and exploration of the qualities and properties of human habitat throughout time. Much of Sauer's published contribution is available in *Land and Life: A Selection from the Writings of Carl Ortwin Sauer* edited by John B. Leighly (1963); a more complete listing is to be found in *Selected Essays 1963-1975: Carl O. Sauer*, edited by Bob Callahan (1981). The whole is in sum an accomplishment which stands alone. His command of the language, and his acquired learning of the human occupance of earth through time, permitted a point of view to be proffered which may only be described as remarkable. At Berkeley this learning took place in an immediate intellectual surround which Sauer helped create for himself. He hired faculty members, decided very largely what courses would be offered (and thereby what was the preferred geography), suggested the locus of fieldwork by his own example, and attracted students to his genre reinforcing the track of an emerging intellectual dynasty. The result was a flowering of sustained excellence. All this has been called the "Berkeley School," a cumbersome approximation to a personal dominion which resulted in intellectual direction.

Sauer's "Morphology" had done much to truncate determinism in American geography. Then emerged a Midwest geography smacking of the region, land-use planning, and economic matters. He characterized this as a period of "Great Retreat." His own studies of the Ozark Highland (1920) and more notably the Pennyroyal (1927) helped to sustain the regional theme which had been gathering momentum since the work of Nathaniel S. Shaler and John Wesley Powell in the 1880s and 1890s. But when the regional concept unfolded, Sauer removed himself from it ("I . . . have objected to the regionalists only because they have been too often short on knowledge and curiosity" *C. O. Sauer to John B. Leighly, July 27, 1943*); then commenced his inquiry into the unspoiled Mexican scapes and points south reaching back to early hominid time.

Historical geography was revivified and made a much more particular field of investigation (largely devoid of "influence") than it had been with Barrows, Brigham, Huntington, or Semple. Sauer adopted a more sustained and far-reaching agenda than these (earlier) workers: he sought the facts and meanings of origins and dispersals in a never-ending series of investigations, in which generations of students were also to participate. Whether this undertaking embraced discipline or whether it was an interdisciplinary scour matters little. For Sauer, "Method is whatever you need to use for the end you are trying to reach" *(C. O. Sauer to "Hal," November 20, 1961)*. It is questionable whether others could share his vision and grasp the interconnectedness of things with a capacity approaching his own. Yet he gave us a literary swath of rare excellence extending over a period of seventy years, a chronologic stretch only occasionally equaled in the history of American geography. And just as Sauer had repudiated the work of some others (nimble and quick folks could be "hyper-adrenals," less quick thinkers could be "compilers," and chairmen, directors, *et al.* might be labeled "administrators"), he was not ardently enthusiastic concerning American geography, finding it too heavily larded with "school-teacher mentality" (for which he largely blamed the Chicago Department). "We are not a science but simply an aggregation of teaching departments" *(C. O. Sauer to P. E. James, March 16, 1940)*. This, in part, doubtless accounts for his usual absence from the annual meetings of the Association of American Geographers. "We already have the factory system instruction and we are too far on the way to the same thing in scholarship to suit me . . ." *(C. O. Sauer to Edgar B. Wilson, October 7, 1944)*. He felt that his fellow geographers had not shown interest in the type of geography that he had developed: "I did try to contribute something to the growth of knowledge and have met with virtually nothing but lack of interest and disdain on the part of American geographers" *(C. O. Sauer to Joseph E. Spencer, September 17, 1943)*. This attitude did seem to change somewhat after he had given a series of five lectures to the American Geographical Society which were published as *Agricultural Origins and Dispersals* which Sauer referred to as "trying to do a geographical Toynbee in five hours" *(C. O. Sauer to George H. T. Kimble, July 17, 1951)*. He sought improvement in American geography and was excited at its occasional prospect. When Whittlesey took a post at Harvard University, Sauer sent him a letter of encouragement concerning the possibility of establishing a first rate department "at the most strategic place in the country":

> I have regarded Bryan's really very honest and sensible observations in the Southwest as some of our best geography of recent years and I was rather jubilant when he went to Harvard, because it seemed to me that it opened the possibility for a rather different sort of geography than has been accepted in the canons of the Great Corn Belt. The acquisition of Blanchard to drill the as yet ragged Continentals became to me therefore the Valley Forge of the Geographic Revolution. And finally you seized the chance to be yourself; I've long wondered when you would transplant yourself and have been somewhat dismayed that you had not done so *(C. O. Sauer to Derwent S. Whittlesey, March 23, 1929)*.

Yet there were but few occasions when his optimism for the cause of geography in North America was thus brightened. He felt that it was necessary to reach out to European geographers, to bring them into American geography. He brought Oskar Schmieder, Wolfgang Panzer, Albrecht Penck, Gottfried Pfeifer, Fritz Bartz, and Karl Pelzer into his own department, in what had initially been a deliberate attempt to create "an institute for European geographers" *(C. O. Sauer to Alfred Hettner, January 7, 1926)*. And Sauer had his students read in the German literature. They read *Die Morphologische Analyse* by Walther Penck—whose opus Sauer regarded as one of the largest contributions to geomorphological science in the first half of the twentieth century—in what was part of his ongoing skepticism of Davisian physiography. (Sauer also encouraged more particularly the import of British, French, and Scandinavian geographers, while he added the Swiss, John Kesseli, to his own department.) He observed that geography was being peddled by slick promoters with "little but good resonance," that crowd action was subverting individual intelligence, that geographic fashion and "Chamber of Commerce geography" had replaced scholarship:

> Right now these boys are all political geographers. Several years ago they were land promoters; before that they were economic geographers. . . . the people who crowd into the thing that is the latest vogue are not likely to be the people who have tenacious curiosity *(C. O. Sauer to Victor E. Monnett, December 14, 1943).*

His own geographic excursus extended itself as an agenda throughout a lifetime, learning begetting further learning, and the whole being forged into a profoundly honed instrument of knowledge. It was a (long) life work. While this spectacle was in progress, American geography swept by in ever more hurried manner, experiencing dramatic change in focus. The result was that Sauer found changed geographies and geographers in juxtaposition to his own work as the years advanced (a circumstance frequently shared by long-lifed geographers). That is in part why David Hooson could write of Sauer, "He was indeed a polymath, recalling great free-ranging scholars of the nineteenth or even earlier centuries. He transcended academic geography,"[4] and why Allen Pred, of the same Berkeley Department as Hooson and Sauer could write of the latter, ". . . listening to Sauer could be extremely disturbing because of his narrow-mindedness. . . . it was not Sauer's refusal to tolerate a more pluralistic view of the nature of geographical inquiry which led to my breaking off our occasional conversations . . . it was rather his narrow-mindedness with respect to the American social scene which forced the rupture."[5] There were bound to be differences of opinion because, throughout Sauer's very active retirement (1957-75), his point of view was so very different from the younger, numerate, model-building generation (and yet others) whom Sauer felt were diffusing "the dreary quantifying and model dreaming that is casting its leaden pall" *(C. O. Sauer to Campbell Pennington, March 15, 1969).* He was saddened by "model builders and system builders and piddlers with formulas of imaginary universals" *(C. O. Sauer to Campbell Pennington, February 4, 1967)* and wanted geographers to return to something more fundamental. He had an urgent concern for the violation of the planet by a technology which had exceeded the intelligence of human command. His feeling was rather well summed in a letter he wrote to Mr. Wickman:

> It is quite possible that our whole western civilization in its modern form, based on every increasing production and consumption, is a violation of natural order which will bring about its collapse. It is possible that the unparalleled malignancy of nationalism in our time is a sickness based on a pathologic industrialization on increasing unbalance between population and resource with increasing failure of resource. *(April 22, 1948).*

Sauer thereby challenged our notion of progress, deplored thoughtless exploitation of resources, placed before geographers (and others) questions concerning the quality of human planetary stewardship, urged study of the occupational inscriptions of humans on the land they occupied, and thereby revealed Antaean qualities of his being.

Sauer's meticulous intellectual exploitation of longevity, dedication, remarkable felicity of expression, and sensitivity to the native values of the rural surround led to a genre of writing that has been without compare in the history of North American geography. It was high art and hardly possible of emulation. Sauer's intimacy with the land (on his own terms) has won new-found appreciation, and historical geographers (and friends of the land, and yet others) have rediscovered the man's remarkable swath initiated in the teens. Since his passing perhaps more sustained attention has been given to this luminary than any other in our history, though there are more of us now to sustain attention than there were when William Morris Davis died in 1934.

This happy volume of essays arranged by Martin Kenzer has been designed to contribute to a further understanding of the life and thought of Carl O. Sauer and thereby our comprehension of the history of geography especially in North America.

Geoffrey J. Martin

Notes

1. "A brief history of the department of geography at the University of Michigan." Anon.

2. Martin, Geoffrey J. "Paradigm Change: A History of Geography in the United States, 1892-1925," *National Geographic Research*, vol. 1 (Spring 1985): 217-235.

3. *Geographical Review*, Volume 16, 1926: 349.

4. *The Origins of Academic Geography in the United States*, edited by B. W. Blouet, Hamden, Connecticut: Archon Books, 1980, page 170.

5. *Recollections of a Revolution*, edited by M. Billinge, D. Gregory, and R. Martin, New York: St. Martin's Press, 1984, pp. 92-93.

Preface

In part, this volume stems from two special sessions on Carl Sauer organized for the 81st Annual Meeting (Detroit, 1985) of the Association of American Geographers (A.A.G.). More importantly, it commemorates the estimable 50th Anniversary of the Association of Pacific Coast Geographers (A.P.C.G.), and it is through the generous support of the A.P.C.G. that this book has been published.

Founded in 1935, the A.P.C.G. has always been and today remains one of the most active organizations of North American geographers. For the past two decades the A.P.C.G. has provided a major segment of the national membership, accounting for approximately 16 percent of all A.A.G. members. The A.P.C.G.'s *Yearbook*—forty-seven volumes strong in 1985—has continually carried some of the highest quality work among west coast geographers. Sauer had little personal involvement with the Association. He was a product of an earlier generation, and his interests, by 1935, were beginning to stray from the core concerns of his colleagues. His influence, however, is evident in each volume of the *Yearbook*. Perusing the contents of back issues, one finds the journal dominated by his students' ongoing research concerns. As an outlet for Pacific Coast geographers, the *Yearbook* may in fact represent, more than any other single collection, Sauer's lasting contribution to a distinctive "school" of American geography. I am certain that it is the desire of the A.P.C.G. membership and of each contributor to this volume, that the present collection will, at least in part, serve as a fitting accolade to one of the best-known figures of twentieth-century geography.

I would like to give thanks to several individuals. James W. Scott, A.P.C.G. *Yearbook* editor, has been of great assistance in this project from the outset. By acting as a managing editor he has eliminated my need to negotiate with the Press at every turn. Without his enthusiasm and support the book may never have come to fruition. I would also like to thank Geoffrey J. Martin for contributing the foreword, and Jo Alexander, managing editor of Oregon State University Press, for her help, including preparation of the index. Finally, I would like to express my gratitude to the Departments of Geography at McMaster University and at the University of Southern Mississippi for their administrative support while I was affiliated with that institution.

Martin S. Kenzer
October 1985

Introduction

This is not the first posthumous tribute to Carl Ortwin Sauer (1889-1975). Since his death there has emerged an eager corps of individuals who have written about the man and his myriad achievements. Some have noted his establishment of and continuous influence on the Berkeley Geography Department (Leighly 1979, Parsons 1979, West 1980, Speth 1981). Others have tried to interpret his writing from a variety of research perspectives (Speth 1977, 1978, Duncan 1980, Kersten 1982, Williams 1983, Entrikin 1984, Kenzer 1985a). One author has written extensively on Sauer's work in and contribution to the subfield of Latin American geography (West 1979, 1981, 1982), while another has identified conspicuous themes in his collected correspondence (Kenzer 1985b). A short while after his death, some of his close friends and colleagues wrote personal recollections to commemorate Sauer's broad intellectual influence both within and beyond the field of geography (e.g., Pfeifer 1975, Leighly 1976, Parsons 1976a, 1976b, Leighly 1978a, 1978b). Similarly, several touching reminiscences appeared on the heels of Sauer's passing which focused on his character and personality (Ballas 1975, Kramer 1975, Stanislawski 1975), whereas two recent essays have provided a more contemplative, reflective picture (Hooson 1981, Hewes 1983).

This, however, is the first collective, monograph-length appreciation of Sauer. To the best of my knowledge, it is also the first contribution of its kind in geography. It speaks to the need for group effort when trying to understand a very complex individual. A person like Carl Sauer must be examined from as many angles as possible. Each of us will appreciate his writings from a number of perspectives; each of us accordingly lends unique insight into the man and his work. It was with this idea in mind that I selected the various contributors to this book. Each author knew Sauer or had deciphered his research from a different background, a different type of training, a different time period; each author has captured a distinctive attribute of Sauer's life and career. Together, the essays should reveal more of the "true" Sauer than any single author may hope to unearth alone.

It is necessary to note that this volume is *not* a *Festschrift*. Properly speaking, a *Festschrift* is an honorific, congratulatory collection of essays. It is

1

presented to an esteemed member of his or her profession during their lifetime. The essays are generally empirical in nature, and usually they focus on the particular research themes of the honored member. Moreover, the essays typically reflect (or *try* to reflect!) the scholarly interests and accomplishments of the individual to whom they were dedicated. This, therefore, is not a *Festschrift*. It is a highly eclectic group of papers brought together for one main purpose: to present a few of the numerous strands which, when woven together, represent the background, career, and intellectual legacy—the biographical elements—of one of America's most important academic geographers.

We can identify at least four distinct categories of writing that purport either to understand or to explain the life of an academician. Traditional biographies tend to uncover the subjective, anecdotal side of a scholar's life. Colleagues' accounts and reminiscences usually point out the more personal aspects of that individual's career. Investigative researchers attempt to unravel a person's intellectual interests and subsequent influence and thus contribute to a third category of writing: professional, disciplinal achievements. Finally, in the case of very prominent members of academia, there is the long-term heritage evident in their graduate students' writings. Students frequently refine, augment, and perpetuate the ideas and insights they were exposed to in graduate school. Their later work, both collectively and independently, often reflects the confluence of methods and research interests encountered in postgraduate studies. In this sense, their work can be seen as a lasting record of their mentor's guidance and represents a fourth category of "biographical" writing. Each of the four types of writings—traditional biographies, personal reminiscenses, scholarly analyses, and students' publications—helps explicate a different side of the person in question. None of the four is fully satisfying alone; each complements the other three. Unfortunately, they rarely exist side-by-side in the same volume. More commonly they are found in disparate publications and, in the case of students' writings, sometimes buried within the journals of divergent academic subdisciplines. This volume is an exception.

The book is divided into five sections. The first section focuses on Sauer's earliest social and intellectual heritage. William Speth's paper introduces and establishes a framework for understanding the remainder of the essays. Speth builds upon his earlier hypothesis that Sauer was grounded in a Goethean world view, but extends this notion to place Sauer's conception of geography squarely in the tradition of German historicism. Sauer's epistemology was so firmly rooted in German ideas and their attendant historical method that there is hardly a thing one can say about the man or his writings without first understanding this fundamental point. The paper by Martin Kenzer extends Speth's findings further by showing the strong continuity between the German, Goethean, historicist tradition of Sauer's father, and the clearly German-American community of Warrenton, Missouri, where Sauer was born and raised. An abiding tie between his father's influence and Sauer's ideas is proposed, and a brief outline is provided of German immigration to Missouri and the intellectual foundation of Central Wesleyan College—the institution where the elder Sauer taught, and the younger Sauer's *alma mater*. Taken together, the two essays suggest a convincing argument for understanding the germination of Sauer's early ideas; they also provide a theoretical framework for interpreting his geographic writings.

The papers by Anne Macpherson and Kent Mathewson examine Sauer's research priorities. Utilizing an array of archival material, Macpherson captures Sauer's initial involvement (and frustration!) with the social sciences and the Berkeley social scientists during the period 1923-33. Few geographers are familiar with this segment of Sauer's career, and fewer still realize the importance of this "outside" contact on his subsequent research agenda. This informative essay details the close associations he formed during his first decade at Berkeley, and it also uncovers the difficulties he encountered trying to promote a major, long-term research project—The Institute of Social Sciences. Some may find cause to argue that Sauer's devotion and commitment to outside affairs helps account for the Berkeley Geography Department's small size and lack of growth during the 1920s and early 1930s. Kent Mathewson's essay is a much wider look at Sauer's research priorities. Beginning with Sauer's earliest milieu in Warrenton, Missouri, and ending with his students' research foci, Mathewson highlights Sauer's affinity for the American South, chronicling how that attraction led Sauer to investigate lands near and beyond the South. The crux of Mathewson's thesis is an explication of Sauer's "antimodernist" leanings, thereby providing us with a necessary insight into the direction Sauer's antimodernism took, in terms of the topics and the peoples he chose to study.

The third section consists of three informal papers by three former graduate students recalling different segments of Sauer's career. Henry Bruman recollects his initial (unforgettable!) encounter with Sauer: It was the mid-1930s, and Sauer had recently completed his first full decade at Berkeley. Sauer was shifting gears in the thirties—still concerned with the abuse of fragile environments, but paying increased attention to culture as a motivating force behind human behavior—and Bruman's account captures a very significant period of his mentor's intellectual evolution. It may be the most "balanced" account of Sauer to date. Homer Aschmann's paper calls attention to a mellow Mr. Sauer. Sauer very well may have been at the height of his writing career during the 1940s—not necessarily in terms of variety or quantity, but some of his best-known and most intellectually challenging essays appeared in print then. It was during this decade that he seemingly made the decision to extend his insatiable curiosity ever backward into the early historical, and then the archeological, record. The Carl Sauer whom Marvin Mikesell writes about in his paper was a mature scholar in his mid-sixties. It was the time of Sauer's so-called retirement, and a period preceding four major works of synthesis (Sauer 1966, 1968, 1971, 1980, the last published posthumously). Mikesell, too, warns that we must combine our efforts if we seek a proper

3

understanding of Sauer, asking *which* Carl Sauer we are talking about. Like all creative people, Sauer's views and interests and ideas changed. Accordingly, it is important not to pigeonhole him as a "this" or "that" type of geographer. From these three essays it is apparent that, although the "Doc" continued to change and to search out and explore new topics and fresh problems, his students all experienced the same exacting training, each imbued with the mentor's perpetual sense of inquisitiveness. The three papers represent a different period of Sauer's teaching career, each tells a slightly different story, and each adds to the growing understanding of Sauer the professor and mentor.

The fourth section of the book contains two edited and previously unpublished lectures by Sauer. Following his "retirement" in 1957, Sauer consented to give the occasional seminar at Berkeley or, less frequently, at several select universities across the country. He offered several such guest presentations at the age of seventy-five. Fortunately, two of these have been recorded, transcribed, introduced, and edited for this collection. One seminar was given at Berkeley and the other at the University of Wisconsin-Madison; both took place in the spring semester of 1964. Though decidedly different in content, both presentations capture an aged Sauer still as sharp as ever, fully articulate, blessed with what may have been geography's most retentive memory, and continually looking to culture and culture history to solve a lifelong quest of understanding human tenure on earth. The first seminar addresses culture and cultural geography as fundamental concerns in the geographer's discovery of place. The second is a broad exposition on the Spanish approach to Caribbean exploitation, particularly with regard to their legendary love of gold. Latin America and cultural geography were of course two of Sauer's favorite research themes; he left an indelible mark on both fields. It was therefore of extreme good fortune that these two transcriptions should survive, and it is indeed an honor to include them in this volume.

The fifth section represents an attempt to identify the mentor's influence on the pupil, and is comprised of papers by two of Sauer's former graduate students: Carl Johannessen and H. L. Sawatzky. Sauer supervised thirty-seven doctoral dissertations from the time he arrived at Berkeley until he signed Sawatzky's thesis in 1967. It is of course spurious to claim that the ideas and writings of these two former students alone represent a total reflection of Sauer's influences, just as it is impractical to consider that Sauer had an all-pervasive impact on each and every student. Moreover, Johannessen and Sawatzky were "late" students—from the fifties and sixties, respectively—so our evaluation of Sauer's influence must be tentative at best. Yet, despite all the disclaimers, it requires very little imagination to see Sauer's active mind present in both of these papers. Both are decidedly broad in scope and speculative in nature. Both are intent to show an abiding relationship between the land and human initiative. Both show a keen appreciation of rural habitats. Johannessen is bent on finding "origins"; Sawatzky looks to his own past as prelude to an interest in prairie land use. The former is focused on the diffusion and domestication of agriculture; the latter attempts to understand the ecological relationship between snow moisture and soil formation. Both essays ring clearly of Sauer's long-term interest in problem solving. Finally, it is important to remember that Sauer was frequently criticized and praised alike because his work was distinctive in its treatment of the obscure. He was clearly outside the mainstream of North American geography, and he seemed to gain strength by remaining outside. In like manner, few North American geographers today would tackle the problems identified in the two papers by Johannessen and Sawatzky. Like the mentor, the student has faithfully identified and solved a problem, not conformed to the flow of the discipline.

The final section contains a single essay by Alvin Urquhart, another Berkeley graduate strongly influenced by Sauer. Urquhart's summation draws our attention to Sauer's comparative method, noting how that method brought Sauer full circle: back to humanity's earliest haunts and, at the same time, back to Sauer's personal encounter with rusticity in the Midwest. Both routes led him ever backward, continuously in search of those elusive "origins." It was the ability to realize the relationship between his own past and human history that gave such vitality and permanency to his writings. As Urquhart suggests, for Sauer the frontier was merely a beginning, but never an end. One is tempted to look at Sauer's research record as an autobiography, an autobiography of the human race.

The thirteen essays that comprise this volume (including Martin's foreword and Sauer's two lectures), though a step in the right direction, are only a beginning. They tell us much about Sauer's past, his development, his interests, and his legacy, but they offer little more than snapshot views of the man. The collection is therefore grist for the biographer's mill, and it is, admittedly, incomplete. Sauer was a critical personage in the evolution of our discipline, and he very well may have been "the dominant intellectual figure of twentieth-century American Geography" (Wynn 1981, p. 11). As such, it is virtually impossible to do full justice to his life and works in the space of so few pages. This fact was understood from the outset.

But despite the volume's inherent failings and its lack of comprehensiveness, it represents an attempt to pay tribute to a gifted man who devoted his energies to his love of cultural diversity and human creativity. In 1952 he defined the "complete geographer," as one who:

> . . . must always be learning about the skills that men employ and about the objects, living and inanimate—total environment—to which such skills are applied. He is interested in discovering related and different patterns of living as they are found over the world—culture areas. These patterns have interest and meaning as we learn how they came into being. The geographer, therefore, properly is engaged in charting the distribution over the earth of the arts and artifacts of man, to learn whence they came and how they spread, what their contexts are in cultural and physical environments (Sauer 1952, p. 1).

Sauer was writing about himself in that passage, for he approximated the "complete geographer" more than any other single individual this century. His interests were catholic and capricious, yet he never lost sight of humanity's hand in altering the native, prehuman landscape. Likewise, he was keenly aware of culture and its manifold effects; a major portion of his academic life was devoted to tracing the effects of culture in various parts of the world. It is in this same spirit that the authors of the essays herein have tried to trace and chart "the arts and artifacts of [one particular] man": Carl Ortwin Sauer.

Literature Cited

Ballas, D. J. "Carl Ortwin Sauer's Life and Work." *Places,* vol. 26, no. 3 (November 1975): 24-31.

Duncan, J. S. "The Superorganic in American Cultural Geography." *Annals,* Association of American Geographers, vol. 70, no. 2 (June 1980): 181-198.

Entrikin, J. N. "Carl O. Sauer, Philosopher in Spite of Himself." *Geographical Review,* vol. 74, no. 4 (October 1984): 387-408.

Hewes, L. "Carl Sauer: A Personal View." *Journal of Geography,* vol. 82, no. 4 (July-August 1983): 140-147.

Hooson, D. "Carl O. Sauer." In *The Origins of Academic Geography in the United States,* edited by B. W. Blouet, pp. 165-174. Hamden, Connecticut: Archon Books, 1981.

Kenzer, M. S. "Milieu and the 'Intellectual Landscape': Carl O. Sauer's Undergraduate Heritage." *Annals,* Association of American Geographers, vol. 75, no. 2 (June 1985a): 258-270.

Kenzer, M. S. "Carl Sauer and the Carl Ortwin Sauer Papers." *History of Geography Newsletter,* no. 5 (December 1985b): in press.

Kersten, E. W. "Sauer and 'Geographic Influences'." *Yearbook,* Association of Pacific Coast Geographers, vol. 44 (1982): 47-73.

Kramer, F. L. "Carl Ortwin Sauer, Geographer (1889-1975)." *Geopub Review,* vol. 1, no. 11 (November 1975): 337-346.

Leighly, J. "Carl Ortwin Sauer, 1889-1975." *Annals,* Association of American Geographers, vol. 66, no. 3 (September 1976): pp. 337-348.

Leighly, J. "Carl Ortwin Sauer, 1889-1975." In *Geographers: Biobibliographical Studies,* edited by T. W. Freeman and P. Pinchemel, pp. 99-108. Vol. 2. London: Mansell, 1978a.

Leighly, J. "Scholar and Colleague: Homage to Carl Sauer." *Yearbook,* Association of Pacific Coast Geographers, vol. 40 (1978b): 117-133.

Leighly, J. "Drifting into Geography in the Twenties." *Annals,* Association of American Geographers, vol. 69, no. 1 (March 1979): 4-9.

Parsons, J. J. "Carl Ortwin Sauer (1889-1975)." *Yearbook,* American Philosophical Society (1975). Philadelphia: A.P.S., 1976a: 163-167.

Parsons, J. J. "Carl Ortwin Sauer, 1889-1975." *Geographical Review,* vol. 66, no. 1 (January 1976b): 83-89.

Parsons, J. J. "The Later Sauer Years." *Annals,* Association of American Geographers, vol. 69, no. 1 (March 1979): 9-15.

Pfeifer, G. "Carl Ortwin Sauer (24.12.1889-18.7.1975)." *Geographische Zeitschrift,* vol. 63, no. 3 (1975): 161-169.

Sauer, C. O. *Agricultural Origins and Dispersals.* New York: American Geographical Society, 1952.

Sauer, C. O. *The Early Spanish Main.* Berkeley and Los Angeles: University of California Press, 1966.

Sauer, C. O. *Northern Mists.* Berkeley and Los Angeles: University of California Press, 1968.

Sauer, C. O. *Sixteenth Century North America: The Land and the People as Seen by the Europeans.* Berkeley and Los Angeles: University of California Press, 1971.

Sauer, C. O. *Seventeenth Century North America: Spanish and French Accounts.* Berkeley: Turtle Island Foundation, 1980.

Speth, W. W. "Carl Ortwin Sauer on Destructive Exploitation." *Biological Conservation,* vol. 11, no. 2 (February 1977): 145-160.

Speth, W. W. "The Anthropogeographic Theory of Franz Boas." *Anthropos,* vol. 73, nos. 1-2 (1978): 1-31.

Speth, W. W. "Berkeley Geography, 1923-33." In *The Origins of Academic Geography in the United States,* edited by B. W. Blouet, pp. 221-244. Hamden, Connecticut: Archon Books, 1981.

Stanislawski, D. "Carl Ortwin Sauer, 1889-1975." *Journal of Geography,* vol. 74, no. 9 (December 1975): 548-554.

West, R. C. *Carl Sauer's Fieldwork in Latin America.* Ann Arbor: University Microfilms International, 1979.

West, R. C. "A Berkeley Perspective on the Study of Latin American Geography in the United States and Canada." In *Studying Latin America: Essays in Honor of Preston E. James,* edited by D. J. Robinson, pp. 135-175. Ann Arbor: University Microfilms International, 1980.

West, R. C. "The Contributions of Carl Sauer to Latin American Geography." In *Geographic Research on Latin America: Benchmark 1980.* Proceedings of the Conference of Latin Americanist Geographers, vol. 8, edited by T. L. Martinson and G. S. Elbow, pp. 8-21. Muncie, Indiana: Conference of Latin Americanist Geographers, 1981.

West, R. C., editor. *Andean Reflections: Letters from Carl O. Sauer. . . .* Boulder, Colorado: Westview Press, 1982.

Williams, M. "'The Apple of my Eye': Carl Sauer and Historical Geography." *Journal of Historical Geography*, vol. 9, no. 1 (January 1983): 1-28.

Wynn, G. "W. F. Ganong, A. H. Clark and the Historical Geography of Maritime Canada." *Acadiensis*, vol. 10, no. 2 (1981): 5-28.

PART ONE

Intellectual Traditions

Historicism:
The Disciplinary World View of
Carl O. Sauer

William W. Speth

Unless he is an annalist or a chronicler the historian communicates a pattern which was invisible to his subjects when they lived it, and unknown to his contemporaries before he detected it.

George Kubler

Carl Sauer leaned on an antecedent historicism, yet he was individually creative. He introduced to American geography an inclusive sense of human time and the concept of culture and thereby fostered the trend in geographic thought away from environmentalistic explanation and toward "greater emphasis on man—his culture and his power to modify nature" (Glacken 1973, p. 131). He pioneered the regional historical geography of Mexico, and in its prosecution "came upon a forgotten prehistoric high culture that extended the archeologic limits of Mesoamerica" (Sauer 1974, p. 285). As the founder of a distinctive school of geography, he furthered the German historical tradition and became as well a "founder of discursivity" in American earth science—an initiator of "an endless possibility of discourse" on land and life by those who would apply his design for inquiry (Foucault 1984, p. 114). These attainments bespeak an inventiveness that grew out of the context of historicism, which, in Croce's view, is "the humanism of modern times" (1941, p. 320).

More broadly, in the execution of his labors Sauer combined the several species of historicism. He construed geography historically and as an art that embraced aesthetic appreciation of the landscape, and he brought to his discipline a considered and total vision of the world about him. In his approach to geographic individualities, Sauer drew upon empirical science in the pursuit of truth and of practical knowledge—he had served in "the Army of the Potomac," and he invoked moral judgment against the destructive use of the earth (Sauer 1952, p. 385). Thus, in his life's work, Sauer unified the

11

aesthetic and the philosophical with the empirical and the ethical strains of historicism; his achievement rests to no small degree on this synthesis.

Sauer's outlook on geography and life was freighted with historical value, revealing unmistakably the stamp of the German historical tradition. Historicism, which made history into a principle, structured and informed his professional thought, particularly during his Berkeley years. In reconstructing the nascent period of Berkeley geography under Sauer (1923-33), I postulated that his philosophy of geography was so deeply rooted in German historicism that it was futile to seek to grasp it fully without concomitant understanding of this phase of Western intellectual experience (Speth 1981, pp. 240-241, fn. 7). Subsequent study of German intellectual history revealed an unexpected complexity and unity in German historical knowledge. More astonishing still was the revelation that Sauer's view of geography constituted an accurate restatement of most of the features of historicism in the framework of American geography. Since all knowledge is relative to what is already known, it cannot, accordingly, be unprejudiced. The perfect continuity between the neo-idealist *Denkstil* and Sauer's style of thought is the subject of this paper. It is an analysis of the configuration of a scholarly tradition of thought and of its application. The guiding questions are two. "What are the attributes, biases, and other distinguishing marks of historicism, and of positivism against which it is aligned?" "What are the specifically historicist signposts in Sauer's conception of geography?" The discussion is organized historically to show the early and persistent antagonism between historicism and the positivist outlook on the world.

Method and Assumptions

(*1*) Whether they know it or not, scientists practice their disciplines in a way that presupposes a "system of opinion," or disciplinary world view. Such features as theory, methodology, and synthesizing philosophy together exhibit a kind of "harmony of illusion," and they can be traced historically from somewhat indefinite proto-ideas (Fleck 1979, pp. 23-28; see also Parsons 1937, pp. 22-24 and Kaplan 1964, pp. 23-24). An example of a scholarly image of the world is historicism, which seeks to comprehend all reality historically, or dynamically (Mandelbaum 1967, pp.22-25; Iggers 1973, pp.456-464). For historicism, understanding a phenomenon is finding its genesis, its prior form, its sources, the significance of its development.

(*2*) Further, a disciplinary world view is an intellectual and intuitive interpretation of scholarly experience, expressing in word and formula the coherence of one's discipline. Once a structurally complete and closed system of opinion has been formed, it offers constant resistance to anything that contradicts it and thus excludes alternative world views. No significant communication can arise between fundamentally different viewpoints. Chronic conflict and mutual aggressiveness are, therefore, earmarks of such philosophical ''-isms'' as historicism and positivism. Sauer was aware of this premise for he wrote that ''the content, organization, and context of geographical ideas merit attention to remind us that there is continuity of interest, that the expression changes with the age, and that *a world view has a point of view*'' (1966, pp. 241-242, emphasis mine).

(*3*) Scholarly visions of the world ''are not conscious inferences from experiences, but orderings of experience, achieved largely unconsciously'' (Whyte 1978, p. 5; see also Jones 1961, pp. 1-19). Insofar as a disciplinary world view functions at a fundamental level in the mind, the individual is generally unaware of it. Gouldner remarks that ''background assumptions'' are ''often internalized in us long before the intellectual age of consent. They are affectively-laden cognitive tools that are developed early in the course of socialization into a particular culture and are built deeply into our character structure'' (1970, p. 32). Once acquired, these mental templates can influence the course of a person's reflection on almost any subject, since they unify all knowledge and inform the practical life. Fleck holds that the ''stylized uniformity'' of an individual's thinking ''is far more powerful than the logical construction of his thinking. . . .certain [logical] connections . . . are considered matters of faith and others of knowledge'' ([1935] 1979, p. 110). Sauer's disciplinary point of view represented but a slight modification of his personal world view (somewhat adjusted for academic constraints), which was acquired before he left Warrenton, Missouri, in 1908. Of the sources of his historicist assumptions, Sauer wrote, ''I was either born or conditioned to look on the Earth and Man historically'' (Sauer 1972).

(*4*) All learning is related to some tradition and society (Fleck [1935] 1979, p. 42; see also Trager 1968, p. 536): ''Cognition is the most socially-conditioned activity of man, and knowledge is the paramount social creation. . . . The very structure of language presents a compelling philosophy characteristic of . . . a community, and even a single word can represent a complex theory.'' Individuals thus assimilate from a group a fund of knowledge, which shapes and directs their thinking; they develop a readiness for perception along preexistent avenues of thought. The society in which Sauer internalized his perspective of idealism was German, expressed in family and community (including three years of schooling in Germany) and in professional contacts after his move to California (Speth 1977, pp. 146-147; Kenzer 1985; also this volume). Historicism was the intellectual and scholarly tradition that ''dominated historical, social, and humanistic studies'' in Germany in the nineteenth and early twentieth centuries, and it dominated the subcultural milieu in which Sauer was raised (Iggers 1973, p. 458). Of this movement Nietzsche himself remarked that ''the Germans are a dangerous people: they are experts at inventing intoxicants. Gothic, rococo . . . *the historical sense and exoticism,* Hegel, Richard Wagner—Leibniz, too . . .'' ([1887] 1918, pp. 182-183, emphasis mine).

(*5*) Underlying frameworks of cognition and emotion are objectified in identifiable configurations of language. Not only characteristic theory, methodology, and presupposition, but also their mental cognates, such as key words and phrases, preferred problems and acceptable solutions, formulas, metaphysical mood, etc., can be determined in a scholar's intellectual product (Lovejoy [1936] 1957, pp. 7-15). One can recognize in Sauer's writings not only the lineage of historicism, which he perpetuated, but the unappreciated fullness of his dependence on this lineage as well. During the early decades of the present century, American geographers were reading the literature of their European cohorts; Sauer alone persistently extracted therefrom the materials of historicist value to fashion a coherent view of his discipline. Whenever he sought orientation in programmatic passages or papers, Sauer consulted the historicist Muse.

The Essential Opposition

Positivism and historicism are phases of the more inclusive philosophies of naturalism and idealism, respectively, that have been mutually opposed through the centuries (de Ruggiero 1932, pp. 568-572). Naturalism apprehends reality as a physical system that is accessible to sense experience. Idealism assumes a subjective view of reality, discovering in humanity a will independent of nature, grounded in a transcendental spirit. Naturalism approaches the world aggressively, as an object to be measured, analyzed, predicted, and controlled. Idealism surrenders the world to will and desire; it is the sphere of freedom, of ideas, of spirit. From the naturalist perspective, the contest between the two views of the world becomes one of the progress and triumph of science from its remote classic origins to the present day. To the idealist, the same quarrel appears as a movement to preserve spiritual values against the threat of mechanistic scientism.

Broadly coextensive with the eighteenth century, the Enlightenment brought naturalism (now expressed in terms of rationalism) to its point of greatest influence in the intellectual development of the West (Pappe 1973, pp. 89-100). During this period, certain tendencies of the European mind reached their culmination. This period also marked the beginning of modern natural science, which presupposed that nature is a closed system of causes and effects (Cassirer 1931, pp. 547-552). Methodologically, it held that there is nothing accidental or arbitrary in nature; everything is ruled by universal and necessary laws. It presupposed further that human activity is implied in and subordinated to nature and that it must, accordingly, be explained by the same universal laws. The movements of the will and the force of values on which the human world is founded are subject to the same universal rules as the movements within the world of physical objects. This mechanical analogy was "emphasized so severely by the philosophy of the Enlightenment that it became finally a complete logical identity" (Cassirer 1931, p. 548; Kolakowski 1968, pp. 1-10). The rationalism of the Enlightenment was challenged and reinterpreted during the romantic reaction only to reappear in positivist guise in the following century. Opposition to the French Enlightenment and its supporters in other European countries took the form of the rejection of its central tenets—rationality, universality, objectivity, and progress (Berlin 1973, pp. 100-112; Willoughby 1966, pp. 1-2). By about 1800, a generation of European thinkers was attacking the perceived errors in the all-leveling natural law of the Encyclopedists. Participants in the romantic revolt attempted:

> . . . to reconcile the demands of the intellect with those of the feelings, reason with imagination, the outer world with the inner life, reality with the ideal, the past with the present, the West with the East. It found literature unequal to the task and called music, painting, all the arts to its aid (Willoughby 1966, p. 9).

The predecessors of the romantics, then, espoused a world vision that was mechanistic, finite, a finished product; to the romantic temperament, it was an unrealistic and inhuman outlook (Brinton 1967, pp. 519-525).

Positivism

Such is the oscillatory background—presented in simplified form—of the clash between naturalism and idealism (see Table 1). Around the middle decades of the nineteenth century, positivism became the prevailing *Weltanschauung* of the European intelligentsia. It was the outlook common to Comte

Table 1. A chronology of European philosophies[a]

	Rationalism	Romanticism	Postitivism	Irrationalism[b]
Germany	1690-1770[c]	1770-1830	1840-1890	1890-1945
France	1640-1810	1820-1850	1850-1890	1880-1970
Great Britain	1620-1790	1790-1840	1830-1890	1890-1970
Western Europe	1650-1790	1790-1840	1840-1890	1890-1970

[a] After Wagar 1977, pp. 207-208. Dates indicate the generalized periods of "greatest efflorescence" for each philosophical perspective. There is no intention to imply rigid dates for the beginning and end of each period; there are premonitions and time lags. Further, given the statures of G. Vico and B. Croce in the growth of historicism, an argument could be made for the inclusion of Italy in the chronology.

[b] Wagar's term is "irrationalism"; comparable labels are "neoromanticism," "antipositivism," "neomysticism," and "anti-intellectualism" (1977, p. 138).

[c] The German *Aufklärung*. Differing from the corresponding movements in England and France, the German Enlightenment was not initiated or sustained by a scientific impulse. Rather, it was immersed in an atmosphere of religiosity and did not focus on political reforms.

and Mill, Darwin and Spencer, Marx and Engels, Haeckel and Pavlov. The ascendency of positivism was simultaneous with the advent of industrialization, with its attendant optimism, and the application of modern science and technology to industrial production. Positivism helped shape this cultural climate into a philosophical program. It was a world view that reflected a civilization newly awakened to the promise of material progress and world betterment (Breisach 1983, pp. 268-285).

Continuing the tradition of Galileo and Newton, the crowning achievement of positivism lay in the progress of all the basic scientific disciplines. It supported the rise of the human sciences in the universities, and through it a revolution in physics, chemistry, and biology took place that recalls the great scientific revolution of the seventeenth century. To the positivist, that which was knowable consisted of the data of sensory experience as verified, analyzed, and organized by the empirical sciences or by a methodology analogous to theirs. Kolakowski writes:

> . . . throughout its history positivism has turned a polemical cutting edge to metaphysical speculation of every kind, and hence against all reflection that either cannot found its conclusions on empirical data or formulates its judgments in such a way that they can never be contradicted by empirical data. Thus, according to the positivists, both the materialist and the spiritualist interpretations of the world make use of terms to which nothing corresponds in experience . . . (1968, pp. 9-10).

Arguing that the world is neither mindlike nor essentially material, positivism insisted on the objectivity of truth. In this sense it contradicted Romanticism and remained basically faithful to the canons of Enlightenment rationalism.

Positivism germinated in France and Britain before Romanticism had run its course (Wagar 1977, pp. 97-106). In both countries the Enlightenment had struck deep roots, and the position of the empirical sciences was established. In addition to momentous political changes, the progress of mathematics and the natural sciences was the most notable fact of life for early nineteenth-century French intellectuals. From the 1770s to the 1820s French *savants* dominated almost every branch of scientific inquiry in western Europe. Their success gave impetus to the French positivist thinkers who later expounded the authority of science and its crucial role in the reconstruction of society. Esteem for science was a tenet of positivism, but it had special recognition among the French.

Britain also provided a congenial climate for the advancement of the sciences. As in France, the scientific spirit was venerated. The new social order was founded on reason and experience and would, it was believed, displace tyrannical tradition. Whereas British positivism was more pragmatic and less systematic than the French form, it too subscribed to the practical ideal of knowledge aimed at world improvement. Utilitarian currents which converged in Bentham further bolstered the positivist spirit of English empiricism. The new positive and scientific era of human thought was adopted widely by scholars who molded it to their own special interests. Spencer and Huxley grafted to the stem of positivism the theory of organic evolution that arose from the researches of Darwin. The social world was "biologized" by Spencer: mechanism, belief in the unity of the cosmos, naturalism, an empiricist theory of knowledge, and religious agnosticism were the leading tenets of his system (Kolakowski 1968, pp. 89-103). Sauer objected to Spencer's evolutionary positivism and censured it as a type of "social scientific monism" (1927, p. 173; see also [1925] 1963, p.326 and [1952] 1963, p. 383). Accordingly, utility and progress were the mainstays of British positivism which acquired an evolutionary tinge when wedded to Darwinian theory.

Romanticism had claimed no vital role in Anglo-French philosophy. Positivism, however, prospered in France and Britain; it harmonized with the rationalist-empiricist mood of national thought and exhibited a tendency to view the romantics as conservative since they granted value to those things that were old, enduring, and firmly rooted in tradition. By contrast, positivism was considered "a far more seditious doctrine" in Germany and occasioned "deep national anxiety" upon its arrival there (Wagar 1977, p. 119). By the close of the Romantic era in German history (ca. 1830-35), German thinkers had created for their audience a robust romanticist self-image. Western Europe was the bleak universe of Descartes and Bacon; Germany alone cultivated the spirit of man. Yet in a few years this older German thought was rejected and the *avant-garde* embraced Western influences, most notably its materialism and empiricism. The characteristic German form of positivism was materialism, less a formal metaphysical stance than a movement in the German lands against the old official—political and religious—idealism. Before the materialist impulse had spent itself, Germany had become transformed (by 1871) into a militaristic and authoritarian state, allied with heavy industry. Science and scholarship in the German-speaking world had exploded between 1830 and 1870; all the great scientists of this period had built on the mathematico-empirical methods perfected in the West. There was the inclination—never fulfilled as in the work of Spencer—to treat the human studies in terms of social physics. These were the decades when the materialist views of Bastian in ethnology and the environmentalist prejudices of Ratzel appeared. The age of Bismarck became an echo of the earlier German *Aufklärung* which had originated in the late seventeenth century. However, the German Enlightenment was restrained: it was the work of Leibniz—not that of Newton or Descartes—which served the Germans at this time.

In summary, the salient theses of positivism are that science is the only valid knowledge and facts the only possible objects of knowledge (scientism); that philosophy does not possess a method different from science (i.e., empirical science of reductionist, deductivist, and realist hue); and that the task of philosophy is to find the general principles common to all the sciences and to use these principles as guides to human conduct and as the basis of social organization (scientifically mediated progress).[1] The great waves of philosophy—naturalistic and idealistic—reached most countries of Europe about the same time but became refracted differently in each one. The uniqueness of the development of philosophy and science in the German-speaking lands rests on the special regard and emphasis given there to the inner life (*Geist*). The paramount feature of German thought is that "it has been anthropocentric in a broad sense, [and] has always taken it to be the job of philosophy to consider the whole embodied person in his environment of nature, history and society" (Sutton 1974, p. 181).

Irrationalism

Perhaps the most encompassing epithet for the view of the world that arose in the twentieth century in Europe is "irrationalism" (Wagar 1977, pp. 137-183). Seen from this angle, irrationalism subsumes historicism by sharing with it an opposition to positivism. It harks back to the romanticist stress on consciousness as will and feeling but goes further in its celebration of unreason. Humans and their world are "fundamentally non-rational, moved by forces beyond the ken of empirical science, and comprehensible (if at all) only by direct intuitive perception or some other non-rational mode of knowing" (Wagar 1977, p. 182). Irrationalism reflects philosophically the problems of the disorganization and decline of Western civilization in our time. It has attacked modernization within capitalism and socialism, it has nurtured and borne witness to fascism, and it has sustained potent forces of reaction in the social sciences, history, and philosophy.[2]

Contrasts in the national expression of irrationalism are most sharp between the Anglo-Saxon countries and those of continental Europe. Political and social problems of the present century were more malevolent in Europe, which was drawn to the interpretations and solutions supplied by the irrationalist outlook. The natural hearths of irrationalism were the German-speaking countries and France, with Britain sharing in its later development. This metaphysical viewpoint was extraordinarily powerful in Germany between 1850 and 1900: proponents include Tönnies, Freud, M. Weber, Meinecke, Windelband, Troeltsch, Spengler, Cassirer, and Mannheim. Analogous to Bentham and Comte in the history of positivism, the great forerunners of irrationalism were (besides Baudelaire in France) Nietzsche and Dilthey, from whom descended nearly all the master themes of later German philosophy.

The German academic elite led the attack on positivism—the *Weltanschauung* of modernity—sensing a threat to its status in the social order from industrialism and social democracy. From the implications and actuality of the dramatic rise to power of industrial leaders on the one hand, and the unionized and politicized workers on the other, the academic intellectuals recoiled. The modernization process, moreover, was incongruent with the German national spirit. Wagar asserts:

> . . . to the mystics, the Reformers, the Kantians, the idealists, the romanticists, the old Germany of Eckhart and Böhme, Herder and Goethe, positivism was thoroughly alien. Although the natural sciences had been converted, more or less, to the positivist world view in the course of the nineteenth century, academic philosophy never made its peace with positivism; history and the social studies preserved links with earlier values, as did law and theology (1977, p. 146).

Such is the background and nature—briefly sketched—of the existential conditions that spawned irrationalism in Europe and released a resurgence of idealist historical thinking in Germany at the close of the last century.

Historicism

By the decade of the 1890s positivist assumptions were undergoing critical review, and novel views of human nature and conduct were being formed against the dominant abstract and *a priori* doctrine (Hughes 1958, pp. 33-66, 183-191). Critics felt that they were rejecting the most pervasive intellectual tenet of their time, and they pitted a "noölogical" perspective against the Spencerian position (Aron [1950] 1964, p. 108). We have seen that positivism failed to gain a hold as firmly in Germany as in France or England. "To a German, " writes H. Stuart Hughes:

> . . . an idealist philosophy was a kind of second nature. . . . Kant . . . remained the dominant formative influence on the German mind. And Kant's contemporaries Goethe and Schiller—the classical writers on whom German schoolboys were nurtured—taught what was usually taken to be a similar lesson. A generation later, and in more dogmatic and eccentric form, Hegel had reinforced the same precepts: like his predecessors, he had built out his doctrine from the idealist premise that the ultimate reality of the universe lay in "spirit" or "idea" rather than in the data of sense perception (1958, p. 183).

Although the idealist revolt was a general phenomenon in Western civilization, it was thus especially strong in Germany. The German thinkers were, above all, acutely aware of the cleavage between their world and the causal-mechanistic patterns and universalism rooted in the outlook of the cultures across the Rhine.

The central idea of historicism goes back to the late eighteenth century and the beginnings of German Romanticism, with Goethe as the genius of the age. Whereas its usage is varied, a substrate of shared meaning that was solidified in a new sense of historical development can be isolated. The fundamental quality of historicism lies in the recognition that all human phenomena are subject to change (Iggers 1973, p. 457). In Germany this rise in historical consciousness "acquired the status of an almost official national philosophy of history," the last great representative of which was Meinecke (Berlin 1972, p. x). Meinecke saw the principles of "development"—the dynamic processes of temporal change—and "individuality"—the concrete uniqueness of that which is historical—as the cornerstone of the historicist world view (Mandelbaum 1967, p. 23). Inasmuch as these two concepts must be grasped in relation to one another, development is the change and growth of an individual whole.

Like many philosophical movements, historicism signified something in crisis, something to be overcome. In this sense, historicism is one of the purest examples of a *Kampfbegriff* (literally, "struggle-concept"). It took shape as the radical view that all forms of cultural life are to be regarded and understood as historical, that is, as developing organic wholes which could be intuited or understood "by a species of direct aquaintance" (Berlin 1972, p. ix). Fundamentally traditionalist, historicism is compounded of a kind of scientific practice, which is contemplative and makes no distinction between what is and is not important, and a philosophy of cultural relativism. The most important achievement of historicism was to bring the older conceptions of history under the emerging historical principle (Schnädelbach 1984, p. 37). History could no longer properly concern itself with uniformly applicable laws or essences (Hegel). Where the human past was concerned, time was reconceptualized—full of vicissitudes, contingently determined, and elapsing in various ways. Applied to humanity the principle of historicity meant that "one can treat as unhistorical only that of whose origin one does not know" (Schnädelbach 1984, p. 38). In the framework of the Enlightenment, reason made humanity human; under the historical principle, it was reduced to the province of the irrational as humanity itself became an historical theme.

The German historical movement thus rejected *a priori* constructions (laws) and sought to observe and understand historical individualities; it emphasized the discovery of connections (organicism) and the description of how things have come to be as they are. *"Understanding observation"* became the fundamental notion in the theory of historical science (Schnädelbach 1984, p. 43). Understanding is an inference from the manifestation of a cultural event to the inner process of which it is an expression. To gain understanding is to retrace the route by which one's subject matter came into being (Schnädelbach 1984, pp. 112, 121-122). Scholars relied on the watchwords of Romantic hermeneutics—empathy, imagination, and sympathetic identification—and they aimed, moreover, at making historical material available and at evaluating it in regard to its trustworthiness. These procedures are basic to historical inquiry and understanding, and Sauer himself spent considerable effort and time on them ([1941a] 1963, pp. 366-367; Parsons 1976, p. 84; West 1982, p. 116). Thus historicism developed as an inclusive mode of reckoning of human life on earth which is approached empirically and interpreted historically.

Excursus: German Science

The transformation of German science, which, by 1830, began to reorganize itself in accordance with the natural sciences, was integrated with the rise of historicism. The classical conception of science was based on universality, necessity, and truth and was accepted from Aristotle to Hegel: science is knowledge of the universal—the causes of the particular, of the necessary—what universally, eternally is, and of truth—demonstrable judgments. Under Descartes, and later Kant, mathematics gained in stature, and classical mechanics became the paradigm of science in general. This triadic conception of science fell victim to the science that developed in the middle decades of the nineteenth century.

Traditionally, science and experience had been regarded as totally different; knowledge of the particular and the contingent (experience) could not be science. This view changed as empirical observation became the criterion of science, displacing the understanding of science as a system or aggregate of knowledge. Empiricism thus emerged as the leading doctrine in German science after Hegel (Schnädelbach 1984, pp. 82-88; de Ruggiero [1921] 1969, pp. 32-33). While it is common to both positivism and historicism, empiricism is not the same thing in the two contexts (Parsons 1937, pp. 476-477). Empiricism of positivist slant is related to general analytical and conceptual schemes which are closed and deterministic. Concrete reality is reflected (illustrated) in the system. On the historicist side, mechanistic determination is rejected in favor of human freedom, and the concrete subject matter is regarded as unique, individual, and historically given. The two forms of empiricism are irreconcilable, since the idealist tendency gives no credence to systems of analytical theory.

Accordingly, the inductivist version of empiricist thought found greater support in the German mind than did the deductivist form. Even in Goethe's time the spirit of inductivism was well established, but nature-philosophy and science had not yet been conceptually separated (Bynum et al. 1981, pp. 292-293). ''Inductive'' became the key word of post-Hegelian philosophy and science, and inductive science came to be regarded as *the* science of reality. At this time, Ranke (1795-1886) pointed the way to inductive research in history in his desire to show ''how it actually happened'' (Goetz 1932, p. 376; see also Gooch [1913] 1952, pp. 96-97). Moreover, the emphasis on observable phenomena in German physics led to a postmechanist, postmaterialist interpretation that can be called descriptivist or phenomenological. Insofar as experience (empirical observation) held sway over theory as the touchstone of scientific status, theories lost their pre-Hegelian importance; they could be discarded or revised under the influence of the inductive process which can lead to synthetic generalization (see Buckle 1872, vol. 3, pp. 290-291). Thus, under the ascendency of empiricism, the traditional static ideal of system was displaced; theories were merely the result of the systematization of experience. The general structural change in science led from a propositionally defined systematic conception of science to a procedurally defined notion of science as process (research).

The type of science to which Sauer subscribed was first cast in the mold of German empiricism; empirical scientific observation was the proto-idea of field observation in his geographic methodology. He was convinced that "geography is first of all knowledge gained by observation, that one orders by reflection and reinspection the things one has been looking at, and that from what one has experienced by intimate sight come comparison and synthesis" (Sauer [1956] 1963, p. 400). His geography was thus "experiential" as well as historical (Sauer [1963] 1981, p. 78). The notion of an interplay between seeing and reflecting was, however, extant before Goethe reached maturity and can be seen in the thought of a lesser known *Aufklärer*, J. M. Chladenius:

> Chladenius's root metaphors are drawn from the act of seeing, not from cognition. At critical junctures he introduces [in his *Allgemeine Geschichtswissenschaft*, Leipzig, 1752] such terms as *Sehepunkt* ("vantage point"), *Zuschauer* ("observer"), *einsehen* ("perceive"), and *anschauen* ("contemplate"). To see means to experience, but not to experience passively. Experience is not mere sense impression etching an indelible picture upon an impassive *tabula rasa*, nor is it a logical category; rather, it is a product of the interaction between sense impression and an active *Gemüthe*, a term that is elusive because it signifies a spiritual mood (Reill 1975, p. 110).

Need one look beyond German historicism to understand the form of Sauer's science—"retrospective" ([1941a] 1963, p. 379), inductive—"never contained by systems" ([1941b] 1963, pp. 1-4), "always genetic" ([1948] 1976, p. 33), and empirical—"a science of observation" ([1956] 1963, pp. 393, 400)?

Further, as a consequence of the renaissance of empiricism around 1800, art assumed a significant role in the human studies. The structure of the historicist outlook was changed when aesthetic considerations were added as a counterweight to its tendency to become excessively scientific (Schnädelbach 1984, p. 95). Sauer defended the step that went beyond science—it was not "mystical"—and suggested that "a good deal of the meaning of area lies beyond scientific regimentation. The best geography has never disregarded the esthetic qualities of landscape, to which we know no approach other than the subjective" ([1925] 1963, p. 344; see also [1956] 1963, pp. 403-404). The configuration of historicism was thus modified during the nineteenth century with the absorption of empirical science (naturalism) and art; both elements appear in Sauer's disciplinary view of the world.

The *Geisteswissenschaften*

The historical movement established its scientific identity in terms of methodology as a means to differentiate itself from an earlier Hegelian view of history and to defend the science of history (and later the human sciences as a whole) against the claim to universality made by the natural sciences. This challenge gave considerable impetus among historians to epistemological reflection on the question of how the human sciences understood themselves. The demarcation of the human from the natural sciences gradually became the paramount problem.

The search for an epistemologically secure foundation for the human sciences was undertaken by various thinkers, most notably Wilhelm Dilthey (1833-1911). Dilthey believed that the special capacities required for the cultural and historical sciences were possessed to a unique degree by the German people. He maintained that it was no accident that the *Geisteswissenschaften,* or human sciences, blossomed in Germany, claiming that "only our nation . . . possesses historical consciousness in the highest sense" (quoted in Plantinga 1980, p. 108).

Starting from the premise that man does not have a nature but only a history, Dilthey joined the effort to establish the autonomy of the human sciences. In the process he built upon "the lasting and even decisive influence of romantic ideas" (Cassirer [1940] 1950, p. 224; see also Brinkmann 1931, pp. 600-602). Emancipated from metaphysics and theology by the work of earlier historians, the "mental sciences" still lacked philosophic justification. To resolve this problem, Dilthey wrote his *Einleitung in die Geisteswissenschaften* (1883). "If he is to be likened to anyone," wrote Carlos Antoni, "he must be seen not as the Kant but as the Bacon of the human sciences. For he sought to give them a method which, if differing from that of the natural sciences insofar as it excluded the idea of a rigid causal connection, remained nonetheless inductive and classificatory" ([1940] 1959, p. 19). From the perspective of the history of science, this marked the beginning of the historical transformation of those disciplines—psychology, law, art, religion, philosophy, etc., and the science of history itself—which were later designated the *Geisteswissenschaften.*[3]

Dilthey sought to provide a philosophy of the whole person, to conceptualize the totality of human psychological and historical reality (Plantinga 1980, p. 35; Schnädelbach 1984, pp. 54-56). His main thesis was that the human sciences must be grounded in the relationship of lived experience, expression, and understanding. In lived experience subject and object are not divided. If one speaks of what one has experienced, one speaks necessarily of the experienced object *and* of the subjective experience. The physical and the psychological are thus mutually dependent. The interpretation of lived experience depends upon the expression of experience: what is experienced and the manner of experiencing it are externalized in "a sensuously perceptible form"—in anthropogeography, in the formation of the cultural landscape (Schnädelbach 1984,

p. 55). We know mental phenomena from within, since they are given to us by our own inner experience; we understand by means of inductive inferences from given expressions. On the other hand, we know natural objects only from without, since they lack subjective qualities of experience. This basal distinction prompted Dilthey's motto: "We explain nature, but we understand mental life" (quoted in Plantinga 1980, p. 33; Aliotta 1914, pp. 206-211).

Dilthey further differentiated the *Naturwissenschaften* from the *Geisteswissenschaften* by characterizing the former as a constructionist system and the latter as a developmental or formative scheme (Makkreel 1975, p. 71). The constructionist plan of the natural sciences is hypothetical from the ground up (deductive bias). In the developmental system of the cultural sciences, hypotheses may be ventured only in the late stages of investigation (inductive bias). The possibility of understanding develops out of a cumulative process of inquiry that culminates in historical interpretation. In this view, history cannot be replaced by natural science, which can master "the infinite multiplicity of things" but which cannot provide a concrete individuality within the sphere of the mind (Aliotta 1914, p. 211).

The interpretation of expressive qualities, furthermore, can be accomplished along two avenues. One can gain understanding by tracing the genesis and development of a particular phenomenon, which was the hallmark of Sauer's approach to geography. Dilthey, however, spoke of another mode of understanding which depends upon perceiving, or intuiting, connections, coherences, structures, or systems within one's object of study. Such terms suggest a unity of the object and are approximations to the meaning of *Zusammenhang,* which may have been Dilthey's favorite word. In the natural sciences such interdependences are posited, whereas within the human order they are immediately given or directly apprehended, since both object and subject participate in fundamentally the same world. As late as 1974, echoes of this concept occur in Sauer's thought when he recalled that ". . . we tried to describe the geographic pattern of human activity and interpret its *meaningful assemblage,* and began to ask how the things seen came to be *together"* ([1974] 1981, p. 282, emphasis mine; see also [1966] 1981, p. 243).

Dilthey's fundamental viewpoint, then, was that human life in its manifold expression and depth is found in history and that man is more than pure reason. He held that the products of human activity (expressions) must be understood and that understanding rests on discovering the meanings of cultural objects and events. His methodological work drew attention to the impossibility of the uncritical application of the scientific apparatus of naturalism to the materials of the human sciences.

Historicism is a term that resists easy use. I have used it to define the dominant philosophical orientation that developed over a period of some three hundred years in the German-speaking lands, realizing that there is among historians and philosophers no consensus as regards its definition and application. Not only has the concept changed according to its stage of development, but it is a many-sided notion which has lent itself readily to the stress and purpose of different students. Nonetheless, Table 2 has been organized to show the characteristic ideas that make up the rival outlooks of positivism and of historicism and, also, the extent to which Sauer participated in the German scholarly tradition. The concepts and phrases were gathered from the literature cited in this study and serve as indicators of the opposing cognitive styles. Since the table depicts two ideal types, it does not represent the point of view of any one author; this must be found in the attempts by individual workers to rationalize their disciplinary frame of reference. Reading across, one sees discontinuity of meaning and purpose and, by imaginative leap, mutual antagonism. Reading down, one finds elements that are related and that logically reinforce each other, perhaps creating an overtone of redundancy. The student who is familiar with Sauer's writings is also familiar with many of the attributes of historicism which are grouped in the right-hand column.

Further Congruences

The striking resemblances between Sauer's conception of geography and the idea of historicism are numerous. Most of the characteristics of historicism have been mentioned or summarized in Table 2. However, some of these features (and others that have not yet been introduced) call for elaboration in order to establish the full force of the argument.

In contrast to the insistence of the *philosophes* on modernity and their optimistic orientation to the future, the Romantics "venerated the origins of things and were fascinated by history and such keys to its secrets as the folksong and the fairy tale" (Craig 1982, p. 191). A century later, the upsurge of neo-Romanticism was signaled "in a burgeoning antimodernity and cultural pessimism that became particularly insistent during the Wilhelmine period" (Craig 1982, p. 204; see also Pribram 1983, p. 218). One of the pioneers of the new Romanticism was W. H. Riehl, who was a geographer and professor at the University of Munich and a critic of the abstract opinions of the *litterateurs*. Riehl was:

. . . convinced that the efflorescence of science and industry represented a misfortune for [his] country, which had caused it to lose its moral and cultural moorings, so that it was now adrift on a sea of relativism and materialism. Only a return to the older, more fundamental values could save its soul.

For Riehl, such values were to be found in the life of the peasantry. This was not a new idea, being rooted in the belief of many of the early Romantics that the peasantry, because of its intimate connection with nature, was the most genuine embodiment of native German culture, that is, of a culture free of artificiality and foreign derivatives and rooted in the life of the people. . . .

25

Table 2. Attributes and symbols of positivism and historicism

Positivism	*Historicism*
Rationalistic	Irrationalistic
Optimistic	Pessimistic
Empiricist, deductive, logically closed	Empiricist, inductive, logically open*
Classification by self-correcting hypotheses	Classification *via* hermeneutic circle
Mechanistic, lawful connections	Organismic, contextual coherence*
General truth and system	Individual truth and phenomenon*
Constructionist, axiomatic, deductive	Developmental, formative, cumulative*
Science of abstraction	Science of actuality*
Objects "copied"	Objects reconstructed*
Science of laws, what always is, being	Science of events, what once was, becoming*
Directly observable present, present-minded	Unobservable past, past-minded*
Formalistic explanation	Genetic description*
Lawful explanation of nature, *Erklären*	Intuitive understanding of history, *Verstehen*∗
Cause-effect	Purpose, motive
Phenomenal	Intelligible
Determinant judgment	Reflective judgment, contemplative*
Value-indifferent natural sciences	Value-related cultural sciences, moralistic*
Connections and structures discerned by means of explanation and construction	Connections and structures immediately given*
Comparative method by means of illustrations	Comparative method as means of understanding*
Mastery of society, social control	Comprehension of the social
Practical motive, serves life	Theoretical (pure) motive, enriches life*
Analytical generalization, as goal of deduction	Synthetic generalization, as goal of induction*

It is to Riehl's writings, which were prolific and widely read, that we can trace the origins of that *Völkisch* strain in German thought that became increasingly strong in the last years of the nineteenth century, as well as the beginning of the antiurban prejudice that was associated with it (Craig 1982, p. 204).

Thus Riehl, who is also to be counted as a founder of *Kulturgeschichte,* led the new Romantics in their declaration of hatred for the city and their idealization of the peasantry. Belief in the wholesomeness of rural existence had been expressed as early as 1797 in *Hermann und Dorothea,* one of Goethe's conservative reactions to the French Revolution. Dilthey, too, was critical of his time and was "alarmed by the rootlessness of modern city life. He claimed that only agricultural occupations with their firm ties to the soil could guarantee

Table 2. Attributes and symbols of positivism and historicism (Cont.)

Positivism	*Historicism*
Necessity, determinism	Freedom, autonomy of will, indeterminism*
Predictive model, closed system	Accidentalist model, contingency
Politically progressive	Politically conservative
"System," a priorism	Experience, observation*
Methodological monism	Methodological relativism, eclecticism*
Reality ordered in terms of universal, timeless, unalterable laws	Reality ordered in terms of relativism, principles of change*
Belief in uniformity, unity of mankind	Appreciation of cultural diversity*
Facts of repetition, recurrent time	Facts of succession, non-recurrent time*
The unique as means to lawful end	The unique as the end in itself, individuality*
Outer expression only	Outer and inner expression, psycho-symbolic
Quantifying, "correlations"	Non-quantifying*
Cosmopolitan, *Gesellschaft*, progressivist	Anti-urban, *Gemeinschaft*, traditionalist*
Emphasis upon similarities, homogeneity	Emphasis upon differences, particularity*
A unified science of all there is	"Beyond science," aesthetic appreciation*
Same cause produces same effect	There are not two equal causes and effects
Objectivity	Subjectivity*
Raison	*Einfühlung*
The *salon* as intellectual center	The university as intellectual center
Scientist	Poet
Enlightenment philosophy	Romantic reaction, counter-Enlightenment
Naturwissenschaften	*Geisteswissenschaften*
Descartes, Voltaire, Buckle	Leibniz, Herder, Droysen (respectively)

* Attributes of historicism that are easily identified in Sauer's major programmatic works (1925, 1941a, 1956) are identified collectively with an asterisk. In these papers hardly a page escapes the pattern of the historicist template, and much of his correspondence shows the same influence.

'steadiness of character'" (Bulhof 1980, pp. 14, 196). That historicism's "most interesting aspect . . . [is] its affirmative attitude toward past and present alien cultures" is an observation that characterizes two generations of German intellectuals (Bulhof 1980, p. 19; cf. Nietzsche [1887] 1918). Sauer's idealist vision of primitive and peasant ways and his disparagement of the consequences of urbanism under capitalism, his confessed "fondness for the provinces and a somewhat emotional attitude that the better world will come through the strengthening of local centers of culture, not from the great capitals" had thus been crystallized by earlier German thinkers (Sauer 1944).

To the positivist historian the possibility of the seemingly accidental, the contingent, poses a threat to positivism's generalizing and systematic approach to the past. What might be called the "accidentalist" model of historical developments—history understood as a series of purely accidental happenings—is the extreme idealist position among developmental models (Bulhof 1980, p. 18; Reill 1975, p. 130). Sauer's approach to history was individualizing and depended upon empathetic identification with the human group under study. He was content "to exercise a disciplined curiosity upon the cultural realities, which are groups of men living together somewhere and at a particular time" (quoted in West 1982, p. 18). When, late in life, Sauer reconstructed the origins of planting he adopted the extreme idealist position: "The transition from collecting to cultivation was by vegetative reproduction and probably began accidentally" (Sauer [1968] 1981, p. 326). As corollary, necessity is not the mother of invention; agriculture "did not originate from a growing or chronic shortage of food" (Sauer 1969, pp. 20-21). Irrationalism asserted itself similarly in the neighboring field of animal husbandry, where Sauer persistently followed the thesis of Eduard Hahn (Leighly 1963, pp. 5-6). Given the mood of historicism, it should come as no surprise that Hahn advanced, and Sauer endorsed, a noneconomic interpretation of domestication. Sauer believed that "animals were domesticated for religious reasons, as symbols of a divinity or as themselves sacred . . . [and] that hunters penned game to assure themselves of meat and thus became stock raisers [was] a myth invented by materialist imagination. [He knew] of no evidence that supports an economic basis of the origin of animal domestication" (Sauer [1970] 1981, pp. 317-318; Sauer 1971). Sauer thus elevated chance over rationalist design and spiritual causation over materialism in his understanding of the beginnings of agriculture. In this matter his historicism is indeterministic, "mystical," and extreme.

The historicist attitude had its precursors in the English pre-romantics, in Rousseau, and in German Pietism. Developing rapidly toward the end of the seventeenth century, Pietism was a reaction against the arid and rigid Lutheranism of the day (Tholfsen 1967, p. 131). Distrustful of doctrine, it embodied an inward and emotional approach to religion and enjoined the individual to cultivate inner piety. A radical transformation of the soul was to follow from an intense faith in the sacrifice of Christ. Prayer and Bible reading rather than the pursuit of dogmatic theology were what mattered most. Although Pietism declined as an organized movement by the middle of the eighteenth century, it left its mark on a generation of German intellectuals, including J. G. Herder who provided historicism's first formal statement and other thinkers who heightened our consciousness of history. These men were aware of their German roots, not only in the Reformation, but in "Pietism and the mystical and visionary movements that preceded it" (Berlin 1972, p. x; Reill 1975, pp. 215-216). The inward-turning tradition of Pietism in Germany strongly reinforced the mood of rebellion against the rationalism and scientism of the *philosophes*. Adherents of the traditional German faith were distinguished by:

> . . . their belief in the unadorned truth, in the power of goodness, in the
> inner light; their contempt for outward forms; their perpetual self-examination;

their obsession with the presence of evil . . . and above all their preoccupation with the life of the spirit which alone liberated men from the bonds of flesh and of nature (Berlin 1976, pp. 151-152).

Pietism left its trace on Sauer, mainly through his family. Sauer's forebears were members of a German Pietistic sect, and his maternal grandfather was a minister of the faith (Leighly 1979, p. 99). His simple way of life—"the idea of a 'vacation' was unthinkable"—and his total dedication to the quiet pursuit of learning —"the contemplation and concentration that are needed for creative work"—imparted to Sauer's existence an introspective dimension (Parsons 1976, p. 84; Sauer [1952] 1963, p. 384). Sauer frequently relied on religious metaphors in his writings, and spoke as a person with an inner awareness of self and of solitude when he remarked, shortly after the entry of America into World War II, that "I suppose it is just going to be that kind of a world, as exciting to the gregarious person as it seems twisted up to the solitary one" (quoted in West 1982, p. 9).

The whole tradition of German philosophy viewed man as essentially active, and moral action was seen as integral to the whole person. "From Hegel onwards," Sutton reminds us, "German philosophy was exercised to find a reconciliation between the apparent deliverances of the natural sciences and the apparent requirements of morality" (1974, p. 181). During the *Aufklärung* moral journals appeared in the German language, one of which "concentrated upon a morality founded upon historical and community values" (Reill 1975, p. 201). The generally recognized leader of the historico-ethical school of economics, Gustav Schmoller, asserted that "the greatest nations, epochs, or men . . . are not those who merely increase production, but those who succeed in propagating moral ideas" (Pribram 1983, p. 216). And Nietzsche castigated "the 'eunuchs' in the harem of history" who, hiding behind the cloak of objectivity, refused to judge the facts (White 1959, p. xxii).

Sauer showed no reluctance to judge the facts and sided with the moralists (Leighly 1963, p. 4; Speth 1977, pp. 154-155). From the 1930s onward Sauer advanced a forceful ethical evaluation of the destructive use of the land under profit economics. "We need not say," he declared, "that it is not for us to cross the threshold of value judgments. We are largely committed to the study of human behavior; it is proper and reasonable that we are troubled about how man has acted for good or evil" ([1956] 1963, p. 404; 1938, p. 494; [1941a] 1963, p. 378). Accordingly, Sauer followed the German tradition of moral activism and was relentless in his criticism of the practices and values that undermined the integrity of the interdependence of life and land.

In the work of Herder (1744-1803), whose impact on the younger Goethe and J. S. Mill was immense, we see the ideas of historicism "at their most vivid, freshly formed and unsystematized . . ." (Tholfsen 1967, p. 132). Herder, who saw diversity and change as the essential reality, sired the related notions of historicism, nationalism, and *Volksgeist* and was, all his life, a "sharp and remorseless critic of the Encyclopaedists" (Berlin 1976, pp. 145, 199). Herder's intellectual debts were numerous; many of his ideas, such as the notion of society as an organism, the rejection of cultural arrogance, etc., had been stated before by others.[4] Yet he originated three doctrines, each of which is closely interwoven with the other two: populism, expressionism, and pluralism (Berlin 1976, p. 153). These ideas suffuse Sauer's thought.

(*1*) Populism refers to the value of belonging to a group (culture) which is not political and is, to some degree, antipolitical and even opposed to nationalism. (His conception of a good society resembles the anarchism of Thoreau or Kropotkin.) To Herder, populism was:

> . . . democratic and peaceful, not only anti-dynastic and anti-elitist, but deeply anti-political, directed against organized power, whether of nations, classes, races, or parties. . . . It is, as a rule, pluralistic, looks on government as an evil, tends . . . to identify "the people" with the poor, the peasants, the common folk, the plebian masses, uncorrupted by wealth or city life; and to this day, animates folk enthusiasts and cultural fanatics, egalitarians and agitators for local autonomy, champions of arts and crafts and of simple life, and innocent utopians of all brands. It is based on belief in loose textures, voluntary associations, natural ties, and is bitterly opposed to armies, bureaucracies, "closed" societies of any sort (Berlin 1976, p. 184).

Certain "paradoxes" that surround Sauer—viz., his political conservatism and distrust of political solutions, his skeptical attitude toward applied geography or employment in the government, his romantic view of rural life and disparagement of urbanism—are clarified, if not resolved, by recognizing the antecedent structure and context of Herder's populism (see Hooson 1981, pp. 166-167). Sauer was among the "progeny of Herder," student of ancient and modern cultures other than his own, defender of the rural provinces of Europe or America, advocate of cultural autonomy (Berlin 1976, pp. 184-185). Most of these themes are familiar to those who are acquainted with Sauer's historical *Weltanschauung*. Less familiar perhaps is Sauer's championing of arts and crafts, which is illustrated by the story he told of an encounter with a "strange native" near Riobamba who was trying to revive the ancient art of rug weaving:

> The product does not quite come off; weaving a bit too uneven, some of the dyed yarn too flat in tone, some of the original designs too bizarre. But this little fellow stirred me and I'm hoping that he will bring back the full lovely art of the colonial days. He is right in all he is trying to do, to revive good hand technique, to use the good old sense of color and pattern, and still to invent something that is more than copying the old (quoted in West 1982, pp. 102-103).

(*2*) We have already encountered in Dilthey's concept of expression a derivative version of Herder's second authentic doctrine. Expressionism means that human activity—art in particular—expresses the personality of the group or individual and that "all the works of men are above all voices speaking . . . not objects detached from their makers" (Berlin 1976, pp. 153, 165ff.). To understand people is to comprehend what *they* mean, intend, or wish to communicate. Language is the primary vehicle for understanding belief and behavior, but one has to feel oneself—*Einfühlen* is Herder's invention—into the collective experience of another people. All genuine expressions of experience are, moreover, valid, unique, and relative; one must not judge one culture by the criteria of another. These strains of Herder's expressionism are repeated in Sauer's time-bound geography: the geographer "needs the ability to see the land with the eyes of its former occupants, from the standpoint of their needs and capacities . . . to place oneself in the position of a member of the cultural group at the time being studied" ([1941a] 1963, p. 362; see also [1966] 1981, p. 259).

(*3*) Herder's third original idea, belief in the value of diversity (pluralism), has been recognized by writers who seek to understand the pattern of Sauer's thought.[5] Pluralism is the belief in the multiplicity, the incommensurability, the variety of different cultures and peoples. Herder feared organization as such and wanted to preserve "what is irregular and unique in life and in art, that which no system [could] wholly contain" (Berlin 1976, p. 183). He opposed any system which imposes the standards and ideals of one group on another, and he saw centralization and state planning as the enemies of freedom and individuality. "Diversity is everything" is the central thesis of his *Yet Another Philosophy of History* (1774) and of most of his early writings as well. Herder was thus one of the earliest antagonists of uniformity as the detractor of life and freedom. Of his role in the intellectual development of the West, Berlin writes:

> All regionalists, all defenders of the local against the universal, all champions of deeply rooted forms of life, both reactionary and progressive, both genuine humanists and obscurantist opponents of scientific advance, owe something, whether they know it or not, to the doctrines which Herder . . . introduced into European thought (Berlin 1976, p. 176).

Sauer's adherence to the value of diversity is revealed precisely in a letter to Joseph Willits, in which he states, "I apparently became a student of man because I enjoy the continuity of pluralism in human history. The exciting thing, I think, is to trace the differences in values by which societies have organized their living and provided for continuity and for change" (1947).

Other striking resemblances between the historicist world images of Herder and of Sauer can only be indicated: preference for the study of cultural origins and the use of the genetic method; reliance on the concepts of development and of organic stages; anthropology as the favored means for understanding human beings and their expressions; true cultural advance as the autochthonous development of a group in its own habitat; colonial subjugation of native populations as criminal and morally odious; and desirability of the qualitative approach of Goethe and Schelling (Berlin 1976, pp. 153-154, 170, 190-193, 213; Nisbet 1970, pp. 22-25, 65-71, 305-310). If one is inclined to forge chains between Sauer and his predecessors, a case could probably be made for stronger links between Herder and Sauer than between Goethe and Sauer; for it was Herder, not Goethe, who originated the "historical sense" in Germany (Stadelmann 1932, p. 327).

At least as early as the period of the Enlightenment, German scholars had employed a formula for reading the future from the past; it was a device that Sauer used for cultural prediction (Reill 1975, p. 214). Past experience, present circumstances, and future possibilities were seen as intimately connected. Historical understanding was the necessary first step to meaningful political and social reform in the German lands. By grasping how the present had been shaped by the past, one could construct a future founded upon concrete reality. Dilthey employed the formula in the context of development to mean that "the present is filled by the past and bears the future within it" (Plantinga 1980, p. 127). The idea blends as well with culture history as with psychology. "Retrospect and prospect," enjoined Sauer, "are different ends of the same sequence. Today is therefore but a point on a line, the development of which may be reconstructed from its beginning and the projection of which may be undertaken into the future" ([1941a] 1963, pp. 360-361; Sauer 1947). Eventually Sauer modified this reasoning; yet during the 1940s (at least) he believed that a retrospective geography was the means by which one could acquire an ability to look ahead (Sauer [1963] 1981, p. 91).

The further congruences that have been noted between the world view of historicism and of Sauer's system of opinion are briefly summarized: aversion to the city and distrust of "progress"; idealization of peasant and rural existence; preference for the accidentalist model in regard to agricultural origins; an introspective quality; expression of moral concernment; incorporation of many of Herder's values and ideas (in particular, but by no means only, the doctrines of populism, expressionism, and diversity); and the use of the past to project the future.

Conclusion

The movement of German thought from the Romantic period to the time of the formation of the *Geisteswissenschaften* is generally called "German idealism." It had no parallel in any of the other countries of Europe. Carl Sauer inherited the *Weltanschauung* of this age and applied it to his discipline; his lodestar became the family of ideas and values known as historicism. In consequence of his early immersion in the German historical tradition, he became a voice of the counter-Enlightenment, fatefully placed in a national culture and professional guild which were heirs to the philosophy of the Enlightenment. His status as an outsider was assured.

The postulate of historicism rests on evidence from intellectual biography (Kenzer 1985, also this volume) and from the history and sociology of knowledge. The pattern of Sauer's cognition was formed in the intimate family and community surroundings of the "Missouri Rhineland." After being repressed somewhat during the Chicago-Ann Arbor period, it was formally set forth in a series of papers by the close of his first decade in Berkeley (Speth 1981, pp. 229-232). The configuration of cognitive and affective features was augmented at Berkeley through cross-fertilization by historical anthropologists, and it persisted to the end of his long life.[6]

There is virtual identity between German historicism and Sauer's outlook on geography, and the manner in which he absorbed it is, at base, no longer a question. As a scholar in the field and archive and as an administrator and teacher, Sauer assimilated new experiences in accordance with his Warrenton *Weltbild*. As he matured and entered old age the details of his idea of the world were enriched but its guiding structure did not alter. The emotional bias and learning that he assimilated in the process of reaching adulthood, his retentive memory, and his later reading of German ethnology and geography combined to reinforce his original perspective (Sauer [1963] 1981, p. 83).

I have suggested that the historicist world view shaped and dominated Sauer's style of geographic thought. The congruences are imposing, both as to number and kind. The sheer density, organic unity, and persistence of historicist features in his writings argue compellingly for the historicist thesis. I doubt, therefore, that the historian of Sauer's ideas can correctly locate their sources in culture history or in his Chicago experiences (Entrikin 1984, 387-390). Culture history is an umbrella term for an historical approach to the social sciences (cf. West 1982, p. 4). It is constrained by the same historicist implications as any one of the social sciences would be if it were approached developmentally. Natural science in its inductive form was fused with historical science in Germany by the time of Sauer's birth; thereby culture history became grounded methodologically in both the genetic method and inductive science (Gooch [1913] 1952, pp. 523-526; cf. Craig 1982, p. 204). I doubt further that Sauer's "model of scientific inquiry" was a result of his postgraduate study in Chicago. While it is true that natural science (empiricism) was part of his geographic training, it was integral to and mediated by his historicist view of the world. I doubt, moreover, that our understanding of the pattern of Sauer's thought is deepened by interpreting it as a "flexible metaphor," which is at times a *Weltanschauung*, possibly of historicist hue (Williams 1983, pp. 1, 5, 14 and 22). This position is inconsistent and tentative.

Finally, there is in Sauer's work a fundamental irony. As he closed his critique of causal geography (positivist anthropogeography) in the essay titled "Recent Developments in Cultural Geography," he implied that the belief in "a single natural law . . .[to] explain the social order"—and by extension by extension the matrix of ideas to which this belief belonged—was a matter of "faith." "Can a field of inquiry," he asked, "surrender its fate to a *Weltanschauung?*" (Sauer 1927, p. 173). Of course not. Nevertheless, Sauer committed his geography to a faith—a predetermined outlook on the world—and thereby brought to American geography a complete historical perspective.

Notes

1. Two types of positivism have been recognized which share the general idea of progress and which shed light on Sauer's view of geography before he moved to California. Evolutionary positivism has a theoretical nature (often materialistically slanted), derives progress from the fields of physics and biology, and looks to Herbert Spencer as its leading representative. Social positivism has a practico-political character, deduces progress from a consideration of society and history, and is exemplified principally in the works of Comte and J. S. Mill. In the social sciences, the pragmatic attitude dovetails with social positivism in its reliance on the method of science, in its penchant for analysis of problematic situations, and in its furtherance of programs to resolve these problems (Kallen 1933, p. 310; Abbagnano 1967, p. 414). Sauer opposed evolutionary positivism which appeared in the guise of the environmentalist definition of geography, and he went through a pragmatic phase during his tenure at the University of Michigan.

2. In his discussion of the organismic (historical) approach to economics, Pribram observes that the sociologist Max Weber shares with many members of the historical school of economics "an outstanding aversion to the capitalist order and its value system" (1983, p. 228). Sauer, too, shared this aversion in criticizing the disruption of symbioses by the commercial economy of the West.

3. The historicist assumption crept into German geography and ethnology, as well. O. Schlüter, who was known to Sauer by the late 1920s, and R. Heine-Geldern, who, as affiliate of the early *Anthropos* group, was indirectly one of Sauer's teachers, aligned their respective disciplines with the *Geisteswissenschaften* (Sauer 1927, Schlüter 1920, Heine-Geldern 1964, p. 416).

4. One of the others was J. G. Hamann, Herder's teacher. The fact that little known contributors (Chladenius) as well as epitomizing symbols (Goethe) of anti-Enlightenment philosophy shared so many of the same ideas depends on their having participated in the same intellectual tradition. They were shaped by and shapers of the same fund of knowledge. Of Hamann, Berlin writes: "As for the famous reversal of values—the triumph of the concrete over the abstract; the sharp turn towards the immediate, the given, the experienced and, above all, away from abstractions, theories, generalizations, and stylized patterns; and the restoration of quality to its old status above quantity, and of the immediate data of the senses to their primacy over the primary qualities of physics—it is in this cause that Hamann made his name" (1976, p. 151). It is in the application of these values to geography that Sauer forged a distinctive view of his field, distinguished, in part, by reliance on a science of actuality, on field experience and intimate seeing, and on description and synthetic generalization.

5. Sauer's pluralism was derived from the German historical outlook, including the ideas of others besides Herder, and cannot easily be understood as a residue of his "studies of plant and animal domestication" (Williams 1983, p. 18). It is more reasonable to conjecture that the problem of the *origin* of cultigens was given by the historicist *Weltanschauung* when it was applied to German geography. Nor can one correctly infer Sauer's "methodological pluralism" from "the inherent diversity of both nature and culture" (Entrikin 1984, p. 389; cf. Williams, 1983, p. 5). Sauer's methodological relativism reflects the German attitude toward the eclectic approach: the historian Droysen stood out resolutely for a "pluralism of methods," and Goethe held that the perception of truth is not the monopoly of any one mode of thinking (Cassirer [1940] 1950, p.

257; Wilkinson and Willoughby 1962, p. 167). It is difficult to believe that problems yield presuppositions and that facts generate methodologies.

6. Even while laboring under a philosophy of practicality at the University of Michigan, the values and vocabulary of historicism asserted themselves in Sauer's thought. Foreshadowing later prejudices in his blueprints for geography, one page in an article on soils geography fairly bristles with historicist watchwords and formulas: "expression," "processes," "derivation," "phenomenon," "genetic factor," "origin," "process of formation," "antecedents," "history," and "how it came to be" (Sauer 1918, p. 85).

Literature Cited

Abbagnano, N. "Positivism." In *The Encyclopedia of Philosophy,* edited by P. Edwards, Vol. 6, pp. 414-419. New York: The Macmillan Company and the Free Press, 1967.

Aliotta, A. *The Idealistic Reaction Against Science.* London: Macmillan and Co., 1914.

Antoni, C. *From History to Sociology: The Transition in German Historical Thinking.* Detroit: Wayne State University Press, 1959 (originally published in 1940).

Aron, R. *German Sociology.* Glencoe, California: The Free Press, 1964 (originally published in 1950).

Berlin, I. "Foreword." In *Historicism: The Rise of a New Historical Outlook,* by F. Meinecke, pp. ix-xvi. London: Routledge & Kegan Paul, 1972.

Berlin, I. "Counter-Enlightenment." In *Dictionary of the History of Ideas,* edited by P. P. Wiener, Vol. 3, pp. 100-112. New York: Charles Scribner's Sons, 1973.

Berlin, I. *Vico and Herder: Two Studies in the History of Ideas.* New York: Viking Press, 1976.

Breisach, E. *Historiography: Ancient, Medieval, and Modern.* Chicago: The University of Chicago Press, 1983.

Brinkmann, C. "Geisteswissenschaften." In *Encyclopaedia of the Social Sciences,* edited by E. Seligman, Vol. 6, pp. 600-602. New York: The Macmillan Company, 1931.

Brinton, C. "Enlightenment." In *The Encyclopedia of Philosophy,* edited by P. Edwards, Vol. 2, pp. 519-525. New York: The Macmillan Company and the Free Press, 1967.

Buckle, H. T. *History of Civilization in England.* Vol. 3. London: Longmans, Green, and Co., 1872.

Bulhof, I. N. *Wilhelm Dilthey: A Hermeneutic Approach to the Study of History and Culture.* The Hague: Martinus Nijhoff, 1980.

Bynum, W. F., et al., editors. "Naturphilosophie." In *Dictionary of the History of Science.* Princeton: Princeton University Press, 1981.

Cassirer, E. "Enlightenment." In *Encyclopaedia of the Social Sciences,* edited by E. Seligman, Vol. 5, pp. 547-552. New York: The Macmillan Company, 1931.

Cassirer, E. *The Problem of Knowledge: Philosophy, Science, and History since Hegel.* New Haven, Connecticut: Yale University Press, 1950 (originally published in 1940).

Craig, G. A. *The Germans.* New York: G. P. Putnam's Sons, 1982.

Croce, B. *History as the Story of Liberty.* London: George Allen and Unwin Limited, 1941.

de Ruggiero, G. "Idealism." In *Encyclopaedia of the Social Sciences,* edited by E. Seligman, Vol. 7, pp. 568-572. New York: The Macmillan Company, 1932.

de Ruggiero, G. *Modern Philosophy.* Westport, Connecticut: Greenwood Press, 1969 (originally published in 1921).

Entrikin, J. N. "Carl O. Sauer, Philosopher in Spite of Himself." *Geographical Review,* 74 (1984): 387-408.

Fleck, L. *Genesis and Development of a Scientific Fact.* Chicago: The University of Chicago Press, 1979 (originally published 1935).

Foucault, M. "What Is an Author?" *The Foucault Reader,* edited by P. Rabinow, pp. 101-120. New York: Pantheon Books, 1984.

Glacken, C. "Environment and Culture." In *Dictionary of the History of Ideas,* edited by P. P. Wiener, Vol. 2, pp. 127-134. New York: Charles Scribner's Sons, 1973.

Goetz, W. "Historiography: Modern Europe." In *Encyclopaedia of the Social Sciences,* edited by E. Seligman, Vol. 7, pp. 374-381. New York: The Macmillan Company, 1932.

Gooch, G. P. *History and Historians in the Nineteenth Century.* 2nd edition revised. London: Longmans, Green and Co., 1952 (originally published in 1913).

Gouldner, A. W. *The Coming Crisis of Western Sociology.* New York: Basic Books, 1970.

Heine-Geldern, R. "One Hundred Years of Ethnological Theory in the German-Speaking Countries: Some Milestones." *Current Anthropology,* 5 (1964): 407-418.

Hooson, D. "Carl O. Sauer." In *The Origins of Academic Geography in the United States,* edited by B. W. Blouet, pp. 165-174. Hamden, Connecticut: Shoe String Press, 1981.

Hughes, H. S. *Consciousness and Society: The Reorientation of European Social Thought 1890-1930.* New York: Alfred A. Knopf, 1958.

Iggers, G. G. "Historicism." In *Dictionary of the History of Ideas,* edited by P. P. Wiener, Vol. 2, pp. 456-464. New York: Charles Scribner's Sons, 1973.

Jones, W. T. *The Romantic Syndrome: Toward a New Method in Cultural Anthropology and History of Ideas.* The Hague: Martinus Nijhoff, 1961.

Kallen, H. M. "Pragmatism." In *Encyclopaedia of the Social Sciences,* edited by E. Seligman, Vol. 12, pp. 307-311. New York: The Macmillan Company, 1933.

Kaplan, A. *The Conduct of Inquiry.* San Francisco: Chandler Publishing Company, 1964.

Kenzer, M. S. "Milieu and the 'Intellectual Landscape': Carl O. Sauer's Undergraduate Heritage." *Annals,* Association of American Geographers, 75 (1985): 258-270.

Kolakowski, L. *The Alienation of Reason: A History of Positivist Thought.* Garden City, New York: Doubleday, 1968.

Kubler, G. *The Shape of Time: Remarks on the History of Things.* New Haven: Yale University Press, 1962 (epigraph quoted from p. 13).

Leighly, J., editor. "Introduction." In *Land and Life: A Selection from the Writings of Carl Ortwin Sauer,* pp. 1-8. Berkeley: University of California Press, 1963.

Leighly, J. "Carl Ortwin Sauer 1889-1975." In *Geographers: Biobibliographical Studies,* edited by T. W. Freeman and P. Pinchemel, Vol. 2, pp. 99-108. London: Mansell, 1979.

Lovejoy, A. O. *The Great Chain of Being: A Study of the History of an Idea.* Cambridge: Harvard University Press, 1957 (originally published in 1936).

Makkreel, R. A. *Dilthey: Philosopher of the Human Studies.* Princeton: Princeton University Press, 1975.

Mandelbaum, M. "Historicism." In *The Encyclopedia of Philosophy,* edited by P. Edwards, Vol. 4, pp. 22-25. New York: The Macmillan Company and the Free Press, 1967.

Nietzsche, F. *The Genealogy of Morals.* New York: Boni and Liveright, 1918 (originally published in 1887).

Nisbet, H. B. *Herder and the Philosophy and History of Science.* Cambridge: The Modern Humanities Research Association, 1970.

Pappe, H. O. "Enlightenment." In *Dictionary of the History of Ideas,* edited by P. P. Wiener, Vol. 2, pp. 89-100. New York: Charles Scribner's Sons, 1973.

Parsons, J. "Carl Ortwin Sauer 1889-1975." *Geographical Review,* 66 (1976): 83-89.

Parsons, T. *The Structure of Social Action.* New York: McGraw-Hill Book Company, 1937.

Plantinga, T. *Historical Understanding in the Thought of Wilhelm Dilthey.* Toronto: University of Toronto Press, 1980.

Pribram, K. *A History of Economic Reasoning.* Baltimore: The Johns Hopkins University Press, 1983.

Reill, P. H. *The German Enlightenment and the Rise of Historicism.* Berkeley: University of California Press, 1975.

Sauer, C. O. "A Soil Classification for Michigan." *Michigan Academy of Science, 20th Annual Report,* (1918): 83-91.

Sauer, C. O. "The Morphology of Landscape." In *Land and Life,* edited by J. Leighly, pp. 315-350. Berkeley: University of California Press, 1963 (originally published in 1925).

Sauer, C. O. "Recent Developments in Cultural Geography." In *Recent Developments in the Social Sciences,* edited by E. C. Hayes, pp. 154-212. Philadelphia: J. B. Lippincott Company, 1927.

Sauer, C. O. "Destructive Exploitation in Modern Colonial Expansion." *Comptes Rendus du Congrès International de Géographie,* Amsterdam, Vol. 2, Sect. 3c, 1938, 494-499.

Sauer, C. O. "Foreword to Historical Geography." In *Land and Life,* edited by J. Leighly, pp. 351-379. Berkeley: University of California Press, 1963 (originally published in 1941).

Sauer, C. O. "The Personality of Mexico." In *Land and Life,* edited by J. Leighly, pp. 104-117. Berkeley: University of California Press, 1963 (originally published in 1941).

Sauer, C. O. to I. Bowman, 11 May 1944, *Sauer Papers,* Bancroft Library, University of California, Berkeley, California.

Sauer, C. O. to J. Willits, 18 December 1947, *Sauer Papers,* Bancroft Library, University of California, Berkeley, California.

Sauer, C. O. "The Seminar as Exploration." *Historical Geography Newsletter,* 6 (1976): 31-34 (originally written in 1948).

Sauer, C. O. "Folkways of Social Science." In *Land and Life,* edited by J. Leighly, pp. 380-388. Berkeley, University of California Press, 1963.

Sauer, C. O. "The Education of a Geographer." In *Land and Life,* edited by J. Leighly, pp. 389-404. Berkeley: University of California Press, 1963 (originally published in 1956).

Sauer, C. O. "Status and Change in the Rural Midwest—A Retrospect." In *Selected Essays 1963-1975,* edited by B. Callahan, pp. 78-91. Berkeley: Turtle Island Foundation, 1981 (originally published in 1963).

Sauer, C. O. "On the Background of Geography in the United States." In *Selected Essays 1963-1975,* edited by B. Callahan, pp. 241-259. Berkeley: Turtle Island Foundation, 1981 (originally published in 1966).

Sauer, C. O. "Human Ecology and Population." In *Selected Essays 1963-1975,* edited by B. Callahan, pp. 319-329. Berkeley: Turtle Island Foundation, 1981 (originally published in 1968).

Sauer, C. O. *Agricultural Origins and Dispersals: The Domestication of Animals and Foodstuffs.* 2nd edition. Cambridge: The M.I.T. Press, 1969 (originally published in 1952).

Sauer, C. O. to W. Speth, 30 June 1971 and 3 March 1972, *Sauer Papers,* Bancroft Library, University of California, Berkeley, California.

Sauer, C. O. "The Fourth Dimension of Geography." In *Selected Essays 1963-1975,* edited by B. Callahan, pp. 279-286. Berkeley: Turtle Island Foundation, 1981 (originally published in 1974).

Schlüter, O. "Die Erdkunde in ihrem Verhältnis zu den Natur- und Geisteswissenschaften." *Geographischer Anzeiger,* 21 (1920): 145-152, 213-218.

Schnädelbach, H. *Philosophy in Germany 1831-1933.* Cambridge: Cambridge University Press, 1984.

Speth, W. W. "Carl Ortwin Sauer on Destructive Exploitation." *Biological Conservation,* 11 (1977): 145-160.

Speth, W. W. "Berkeley Geography, 1923-33." In *The Origins of Academic Geography in the United States,* edited by B. W. Blouet, pp. 221-244. Hamden, Connecticut: Shoe String Press, 1981.

Stadelmann, R. "Herder, Johann Gottfried von (1744-1803)." In *The Encyclopaedia of the Social Sciences,* edited by E. Seligman, Vol. 7, pp. 327-328. New York: The Macmillan Company, 1932.

Sutton, C. *The German Tradition in Philosophy.* New York: Crane, Russak & Company, 1974.

Tholfsen, T. R. *Historical Thinking: An Introduction.* New York: Harper & Row, 1967.

Trager, G. L. "Whorf, Benjamin L." In *The International Encyclopedia of the Social Sciences,* edited by D. Sills, Vol. 16, pp. 536-538. New York: The Macmillan Company and the Free Press, 1968.

Wagar, W. W. *World Views: A Study in Comparative History.* Hinsdale, Illinios: The Dryden Press, 1977.

West, R. C., editor. *Andean Reflections: Letters from Carl O. Sauer While on a South American Trip Under a Grant from the Rockefeller Foundation, 1942.* Dellplain Latin American Studies, No. 11. Boulder, Colorado: Westview Press, 1982.

White, H. V. "On History and Historicisms." In *From History to Sociology: The Transition in German Historical Thinking,* by C. Antoni, pp. xv-xxviii. Detroit: Wayne State University Press, 1959.

Whyte, L. L. *The Unconscious before Freud.* New York: St. Martin's Press, 1978.

Wilkinson, E. and L. Willoughby. *Goethe: Poet and Thinker.* New York: Barnes & Noble, 1962.

Williams, M. "'The Apple of my Eye': Carl Sauer and Historical Geography." *Journal of Historical Geography,* 9 (1983): 1-28.

Willoughby, L. *The Romantic Movement in Germany.* New York: Russell & Russell, 1966.

Like Father, Like Son:
William Albert and Carl Ortwin Sauer

Martin S. Kenzer

The son accompanies his father everywhere—to the field and to the garden, to the shop and to the counting house, to the forest and to the meadow . . . in all the work his father's trade or calling involves. Questions upon questions come from the lips of the boy thirsting for knowledge . . .

Friedrich Frobel

In a related paper I called attention to Carl Sauer's undergraduate intellectual heritage by briefly outlining a few of the activities and three of the individuals most influential in Sauer's life at that time (Kenzer 1985a). At present, I wish to elaborate on a number of the points merely mentioned in that earlier paper. Specifically, I intend to show (a) that Sauer's hometown *alma mater*—Central Wesleyan College (C.W.C.)—was a direct outgrowth of the massive immigration of German "forty-eighters" fleeing adverse conditions in the Fatherland; (b) that the institution, like the rest of the northern Ozarks, was wholly Germanic in every respect; (c) that Sauer's father— (William) Albert Sauer—was the embodiment of the Central Wesleyan *Weltbild;* and (d) that W. A. Sauer and C. O. Sauer's respective world views were, at base, one and the same.

The paper will begin with a general overview of nineteenth-century German emigration to America. This is followed by the peculiar circumstances of German-American immigration and settlement in frontier Missouri, with a quick look at the Sauer family in the context of that society. We shall then briefly examine the C.W.C. milieu, with particular reference to Sauer's father as the epitome of and spokesman for the institution's predominant philosophy of learning. In this section I focus attention on a sample of the elder Sauer's writings, in an attempt to show the relationship between his world view and his son's interest in historical geography. The paper ends by indicating the influence of Johann Goethe, most notably his approach to scientific explanation. Here I endeavor to show how Carl Sauer's philosophy of learning reflects a Goethean world view and how that influence can be traced back to Sauer's earliest milieu. In sum, this paper is designed to accomplish two parallel objectives: to indicate the context of and the circumstances surrounding Sauer's upbringing in Warrenton, and to draw the connection between his hometown intellectual environment and his Goethean, experiential conception of human geography.

Waves of German Emigration

The nineteenth century witnessed the great, unprecedented, continual movement of people and ideas across the Atlantic to America. The Germans, however, entered the flow of that migration in distinct stages. The bulk of their emigration occurred between the years 1816 and 1885. Within that seventy-year span, there were four waves of emigration, each wave corresponding to specific social or political conditions (both in Germany and the United States).[1] The initial departure, beginning about 1816-17, corresponded with an extremely unsettled social climate in southwestern Germany and concomitant years of exceedingly severe harvests. A second wave, much larger than the first, began about 1830 and lasted until the mid-1840s. This second exodus seems to reflect a period of adjustment among the numerous German states to the changing political geography of Europe; it was, in the main, a time of prolonged internal conflict following the Napoleonic Wars and general tumultuous conditions for the coalescing German Empire (Blackbourn 1984). A third nineteenth-century wave of German emigration began in 1848—the year of the so-called "German Revolution"—and lasted well into the middle of the following decade. While small in terms of the numbers of actual emigrants arriving in the United States (Rippley 1976, p. 51), this third wave of emigration was, by far, the most important of the century.

The first two waves of emigrants were composed mainly of peasants, farmers, and the poor. The revolution of 1848, however, has been called "the revolution of intellectuals" (Namier 1944) and those who left Germany in its aftermath were among its most well-educated and highly cultivated citizens. According to one writer, this was the exodus of Germany's "men of distinction"—a generation of scholars and influential orators who were "the true heirs of Kant, Fichte, and Hegel in their devotion to freedom of thought and belief . . ." (Wittke 1967, p. 192; also see Wittke 1952 and 1973, pp. 59-74). Among the more prominent intellectuals who found reason to flee Germany at this time were men like Carl Schurz (1829-1906), Friedrich Hecker (1811-1881), Karl Heinzen (1809-1880), Wilhelm Rapp (1828-1907), Frederick Niedringhaus (1837-1922), Franz Sigel (1824-1902), Gustav Struve (1805-1870), and many other Germans who subsequently became well-known American citizens. "There can be no doubt" writes Wittke, "that the arrival of the 'Forty-eighters' brought about a unique intellectual and cultural renaissance among the Germans in America, and in two centuries of German immigration no other group made such an impact upon the United States as the few thousand political refugees of 1848'' (Wittke 1973, p. 37).

In truth, the complete failure of the 1848 revolution in Germany was most likely the final impetus that many required before they would actually decide to leave their homeland (see Billigmeier 1974, p. 79; Wyman 1984, pp. 57-63). And indeed the greatest percentage of those leaving during this third stream of emigration did not in fact depart in 1848. Rather, they waited until the early- to mid-1850s, well after the hostilities at home had subsided (Wittke 1952, pp. 3, 43). Many of these "revolutionaries" were reluctant to abandon the *Vaterland.* When they finally did leave, it was with the intent of returning to Germany when the European political climate changed for the better.[2] In part, this would account for their penchant for inhabiting the frontier portions of the western United States and clinging to an enduring resistance to speaking English. They were a proud people, a displaced group of intellectuals, and they emigrated with a desire to create small, distinctive German states in the New World. As we shall discover below, Central Wesleyan College was one result of this third wave of German emigration.

A fourth and final wave of nineteenth-century German emigration lasted from about 1871 until 1885. Though important in regard to the large numbers of *Auswanderer,* this last flood of German-American immigrants was relatively insignificant in terms of its lasting importance to the United States. Remarkably, this fourth wave corresponds generally with a period of financial well-being in Germany. Indeed, after German unification (1870) the new nation saw at least twenty years of growth and prosperity (Rippley 1976, pp. 81-83). This final exodus represents migrants heading *toward* something they perceived as better, instead of *away* from something familiar that they feared.

Location of German-Americans

The location of German-Americans during the nineteenth century is a revealing story. Although the arriving Germans settled in all parts of the United States,[3] there was a curious tendency for each new immigrant wave to move farther west with the ever-changing frontier. Thus Faust (1969) has noted that it is important to distinguish between American-born, second-generation German-Americans, on the one hand, and their first-generation German parents who left Europe to settle in the United States, on the other. The former tended to assimilate more readily and to live in heavily populated urban areas. The newcomers, however, had a tendency to head immediately for the then-current frontier of settlement, and to isolate themselves from the rest of society, often intending to create a virtual "Germany in America" (O'Connor 1968, pp. 67-97; Hawgood 1970, pp. 93-226).

By the middle of the nineteenth century, when German immigration was nearing its peak, the American frontier was synonymous with the Mississippi Valley (Turner 1920, pp. 136-139). Nor surprisingly, this is where we find the greatest concentration of budding German-American communities during this period (Faust 1969, vol. I, pp. 432-467). Given Germany's existing social environment, it is apparent why the Germans on the American frontier would strive to create discrete, German-like settlements. There was a profound desire to distinguish themselves from their Old World relations. In addition, there was open concern to establish a sense of identity distinct from their assimilated compatriots in the eastern states of America. As one astute writer has noted:

> Conditions prevailing in both Germany and America at the time favored the rise of a Germanism which in the isolation of the Western frontier frequently sought to further its cause independent not only of American influences but also of contacts with the older German culture of the East. We are dealing with the rise of a unique German civilization which, sometimes divided against itself, did not begin to integrate with American culture until the national crisis of the Civil War (Schneider 1939, p. v, also see pp. 27-30).

For the purposes of the present study, however, we shall ignore the settlement frontier of the entire American Middle West—"the German-American heartland" (O'Connor 1968, p. 5; also see Johnson 1951)—and instead focus our attention on the state of Missouri alone, where Carl Sauer was born and raised.

Mid-Nineteenth-Century German Settlement in Missouri

Although German-speaking peoples settled in what is presently Missouri as early as the 1770s (Ellis 1929, pp. 68-69), it was not until the late 1820s and the early to mid-1830s that the state's German population began to grow in sizable numbers. Some claim that the flood of Germans into Missouri was primarily due to the influence of Gottfried Duden (e.g., Hawgood 1970, pp. 23-24; O'Connor 1968, pp. 68-70; Gerlach 1976b, pp. 14-15). Duden came to America "to become a farmer." He found his way to St. Louis in 1824 and soon thereafter migrated farther west, to what is today Warren County, Missouri, immediately north of the Missouri River in the northeastern Ozarks (Kargau 1900; cf. Schneider 1939, pp. 15-16). Duden's exaggerated reports of Warren (originally a part of Montgomery) County attracted considerable attention in Germany (Duden 1829), and he was undoubtedly guilty of convincing a large number of German immigrants to make their way to the newly constituted state of Missouri:

> His [Duden's] skillful pen mingled fact and fiction, interwove experience and imagination, pictured the freedom of the forest and of democratic

43

institutions in contrast with the social restrictions and political embarrassments of Europe. Many thousands of Germans pondered over his book and enthused over its sympathetic glow. Innumerable resolutions were made to cross the ocean and build for the present and succeeding generations happy homes on the far-famed Missouri (Faust 1969, vol. 1, p. 441).

Duden was apparently responsible for the *Giessener Gesellschaft*'s decision to migrate to Missouri, as they had originally intended to settle in Arkansas (Hawgood 1970, pp. 109-110). Likewise, as a result of what he had written, the *Deutsche Gesellschaft* of Philadelphia probably chose a similar portion of Missouri for the establishment of its "colony" in Hermann (see Bek 1907). On the other hand, it should be noted that Missouri was also the focus of "at least a dozen other German-language travel books . . . by 1830" (Gerlach 1976a, p. 28), so Duden alone cannot be held accountable for the thousands upon thousands of Germans who chose to make Missouri their destination. Whatever his ultimate legacy may have been to the "Show Me" state, it is a fact that the German population grew steadily after the 1829 publication of Duden's infamous book, and "during the third decade of the nineteenth century Missouri became the most favored location for German settlers in the West" (Schneider 1939, p. 19). As early as 1837, the number of Germans moving into distant Missouri was exceeded only by those migrating to Ohio to the east (see Gerlach 1976a, p. 35).

By the 1850s, segments of Missouri were beginning to resemble authentic German "colonies" in terms of their ethnic and social characteristics. Ellis reports that in 1850 the number of Missouri residents who were born in Germany had already exceeded 44,000 (Ellis 1929, p. 126), and Missouri was rapidly achieving the German flavor it would bear for the next one hundred years. In 1855, Oskar Falk described what he considered a typical German settlement in Missouri. It is worth relating his portrayal of these communities to suggest the degree of "Germanness" one might have encountered in the predominantly German-American sections of the state:

> The German settlements in the West are remarkable for their completely German appearance and their purely German atmosphere. While the German farmer in Pennsylvania is more accustomed to Anglo-American ways, and has sacrificed his native language, or half of it at least, on the altar of his new Fatherland, the German settlements in the West have preserved their native colouring pure and unmixed. You think that you are in a village in Germany when you set foot in one of these settlements. The architecture of the houses, owing of course to differences in climate, is a little different, but the household furnishings, the family customs, the style and method of plowing, sowing and harvesting all remind one of Germany (quoted in Hawgood 1970, p. 130).

But early 1850 was really only the beginning of the massive German influx into Missouri. As noted earlier, most revolutionary Germans emigrated to the United States several years after the 1848 uprising, and those who made Missouri their destination during this third wave began to leave Germany predominantly in groups. By mid-century approximately three-fourths of the German immigrants in St. Charles and Warren County (Sauer's home) could be traced back to a thirty-mile radius within Westphalia, usually to the same few villages (Kamphoefner 1978, 1982). In other words, by the 1850s, German-American communities in Missouri were becoming more than haphazard, circumstantial agglomerations of German immigrants: they were often transplanted German villages on American soil.

Between 1850 and 1860, German immigrants virtually poured into Missouri. It has been observed that they decided not to settle ubiquitously across the state, but tended to concentrate instead in the fertile, border areas of the Ozarks, notably along the Missouri and Mississippi rivers (e.g., Gerlach 1976b, pp. 11-16). Turner noted that the Germans in Pennsylvania preferred to establish their communities in and near the limestone areas (Turner 1901), presumably due to the experiences of particular groups who were familiar with a karst-like topography back home. This pattern seemed to hold true all across the country (see Faust 1969, vol. 2, pp. 34-37), and indeed much of Missouri under German settlement was in the limestone portions of the state as well (e.g., Collier 1953, p. 18).

As mentioned earlier, a large number of those entering Missouri during this period headed immediately for the Missouri River area where Duden had already attracted thousands with his eloquent pen and his vivid imagination. So heavily concentrated with German-speaking peoples was the district to become that it would subsequently be called the "Missouri Rhineland" (Rafferty 1982). Faust notes that "this [area] was destined to become the centre of the most widespread settlement of Germans west of the Mississippi" (Faust 1969, vol. 1, p. 444), while another writer calculated that, by 1860, the Germans represented "probably more than two-thirds of the entire foreign population" of the northern Missouri Ozarks (Schultz 1937, p. 75). Faust specifically elaborates on the strong degree of German influence in this riverine portion of the state:

> On both sides of the Missouri River, from its mouth, a little to the north of St. Louis, upward a distance of about 125 miles, all is German territory. In all towns from St. Louis to Jefferson City, such as St. Charles, Washington, Hermann, Warrenton [Sauer's home town], Boonsville, and even beyond and including Kansas City, the Germans are very numerous, generally constituting over one half of the population (Faust 1969, vol. 1, p. 444).

In terms of the county where Sauer was born, he contends that, by 1870, *"On the north side of the River the Germans numbered nine tenths of the population in Warren County"* (Faust 1969, vol. 1, p. 444, emphasis added).

I doubt it is an overstatement to say that Missouri's German population grew at an unprecedented rate between 1850 and 1870. Whereas there were probably less than 45,000 Germans living in the state in 1850 who were born in Germany, the number of native-born Germans climbed to nearly 114,000 by 1870 (Ellis 1929, pp. 126-129 and 152); and in our study area of the Northern Ozark Border, the "foreign-born population . . . was greater, proportionally, than in the state as a whole" (Collier 1953, p. 50). Hence, while Faust's estimate (above) that Warren County was 90 percent German in 1870 may be slightly exaggerated, it is certainly no stretch of the truth to say that the region was unquestionably Germanic in character. In fact, it has been found that the population of some towns in this "Rhineland" region were actually 100 percent German at the time (e.g., Johnson 1951, p. 13). Even today, many towns in the Missouri Ozarks exhibit considerable evidence of their deep German heritage (e.g., Gerlach 1976a, especially pp. 59-109; also see Gerlach 1973). Some communities, like Hermann, Missouri, for instance, are still distinctively Germanic in both appearance and culture (see Roueché 1982, pp. 59-86); at least as late as the 1930s, German continued to function as the preferred language in Hermann (Bratton and Langendoerfer 1931).

Central Wesleyan College

As noted above, Central Wesleyan College was a direct outgrowth of the third wave of nineteenth-century German overseas emigration. It was a small, religious, bilingual college (Fig. 1) isolated from the mainstream of American society in the Northern Border region of the Missouri Ozarks. The C.W.C. faculty were predominantly Germanic, many trained entirely in Germany. Even as late as 1908—the year Carl Sauer graduated from the institution—the Central Wesleyan teaching staff was still dominated by men and women of German ancestry.[4] There was a strong allegiance to German culture at the college and the C.W.C. *Weltbild* was heavily influenced by the writings of Johann Goethe and his followers (Kenzer 1985a; cf. Speth 1981; Williams 1983).

The institution began in Quincy, Illinois—a town, like the general St. Louis area, saturated with German culture and philosophy, understandably the home of the so-called St. Louis School of (German idealistic) Thought[5] —and the first instructors at C.W.C. were refugees of the 1848 German Revolution (Haselmayer 1960, 1964; cf. Billigmeier 1974, pp. 84-85). The "forty-eighters," writes Wittke, "were well educated, in the best German classical tradition. . . . They were," he elaborates, "the liberal heirs of the liberal traditions of Gotthold Lessing, Ludwig Feuerback, and Ludwig Büchner, and of Germany's golden age of liberalism and rationalism" (Wittke 1973, pp. 73-74). More importantly, however, they were also the bearers of the notion of *Wissenschaft* which would ultimately triumph over an earlier generation of *Naturphilosophen* with their thoroughly romantic view of humanity and nature (McClelland 1980, pp. 151-189). Basically, the forty-eighters were scientists who relied on induction and empirical observations; scientists totally immersed in a world of experience, who bore an unrelenting

Typical issue of *The College Star* showing the bilingual nature of Central Wesleyan College. This is the cover of the November 1898 issue.

respect for history (Mandelbaum 1971, particularly pp. 41-138; see also Mendelsohn 1964, especially pp. 39-40). In large part, the forty-eighters who started C.W.C. were schooled entirely in German universities which soon thereafter became the "model" for most "institutions of higher education in the western world" (O'Boyle 1983, p. 3).

German academic learning, we should recall, was highly prized in nineteenth-century America. When thorough German training was unavailable at home, thousands upon thousands of young Americans were sent to Germany to avail themselves of a truly first-class education (Herbst 1965, pp. 1-22). When a traditional German education *was* possible in the United States, the local citizenry usually would opt to send their children to these institutions. In fact, German pedagogical standards became so popular and made such marked inroads into American institutions that by the turn of the century they were considered the ideal and they were to have the greatest domestic influence, exceeding both the French and the British systems (Hinsdale 1899, particularly pp. 603-629; see also Viereck 1978). This would explain, in part, why Central Wesleyan had little trouble attracting students from a wide radius, despite its remote location (Kenzer 1985a, p. 261).

The Sauer Family

One of the professors at C.W.C. was Carl Sauer's father, (William) Albert Sauer (1844-1918). W. A. Sauer emigrated to the United States from the south of Germany in 1865, apparently due to his poor health. Little is known about the elder Sauer before he settled in Warrenton, or indeed even why he eventually chose to come to Missouri. Immediately upon his arrival in America he stayed with relatives in Pennsylvania and Michigan, worked for a short time as an organist in Bloomington, Illinois, and soon thereafter migrated to Warrenton to join the Central Wesleyan faculty (*Warrenton Banner* 1918). His stay in Warrenton began in 1866, but he returned to Germany in 1868 to attend to his ailing mother. He then journeyed back to Warrenton in 1875 where he married Rosetta J. Vosholl (1855-1942, sister of Henry Vosholl, see Kenzer 1985a) three years later (*Central Wesleyan Star* 1918).

In 1881, (William) Albert and Rosetta Sauer had their first child, (Henry) Albert. Like his father, (Henry) Albert Sauer (1881-1936) availed himself of a higher education and attended C.W.C. Unlike his father, however, H. Albert simply was not cut out for an academic future. He first became a skilled machinist and later turned to farming. In 1906 he married Nellie Paul (1880-1962) of Alton, Illinois, and they remained together, childless, until his early death in 1936 (*Warrenton Banner* 1936b). It seems as though H. Albert Sauer may have inherited his father's penchant for poor health. The younger Sauer was frequently ill and eventually died following a long-term "lymphatic infection'' (*Warrenton Banner* 1936a). Beyond these few scant biographical details very little is known of the Sauer's first son.[6]

In the 1880s and 1890s the Sauers held a moderately respectable position in the small community of Warrenton. They were not rich, nor did they own a great deal of land. W. Albert Sauer, however, was head of the Department of Music at C.W.C. and a man of high profile within the local academic community. As a college professor, he enjoyed a measure of prestige similar to the so-called "German Mandarins" of the Fatherland during this period (see, for example, Ringer 1969, Paul 1984). Professors were highly revered and they carried an air of distinction, an honor accorded them by the mere fact that they were academicians in an age when higher education was still of great repute. It has been noted that William Sauer had several offers of employment from a number of more prestigious universities, but he declined each one due to a strong commitment to Central Wesleyan (FitzSimmons 1984) and a similar devotion to the local German-American community at large.

Mrs. Sauer, a teacher for several years prior to her marriage, was likewise a highly respected citizen of Warrenton. Affectionately known locally as "Mother Sauer" (*Warrenton Banner* 1942), she was hardly the inconspicuous housewife we might assume or expect under similar circumstances.[7] In short, the Sauer family was a step above the average contemporary German-American family of Warren County. At the turn of the century, for example, when the Sauers returned from a three-year stay in Germany, the local newspaper hailed their arrival for all Warrentonians to read, announcing proudly that "Prof. A. W. [*sic*] Sauer and his estimable family" were now back home (*Warrenton Herald* 1901).

The general cultural climate at Central Wesleyan retained a strong spiritual link to the common German heritage. The Sauer family (and particularly W. Albert Sauer) were no exception. While undoubtedly American in their allegiance, their sympathies remained with the homeland till the end. Neither William Sauer, Carl Sauer, nor C.W.C. can seriously be examined outside of this strong German context. In countless letters between Carl Sauer and his parents the subject of Germany and German culture would arise. Until 1918, all letters among family members were in the German language. At the outbreak of World War I, the correspondence understandably turned to the German situation and little else seemed to matter to the Sauers.

Little acculturation had taken place in Warrenton, Missouri, in the 1880s and 1890s. The entire community of Warrenton was decidedly steeped in German culture, to the extent that anyone born into this milieu would have a strong preponderance to mirror the local, predominant *Weltanschauung*. In 1889, when the Sauers bore their second child, Carl Ortwin, there was of course no way of predicting, precisely, what his interests might be or what sort of person Carl would become. There was no way of telling whether he would be inclined to pursue a practical vocation like his older brother, or whether his verve would carry him into an academic career such as his father and his father's father enjoyed (see *Beinsteiner Heimatblatt* 1920). No one could forecast accurately that the youngest Sauer would amount to anything at all; his precociousness (see Kenzer 1985b) would not have been evident at so early an age. However, given the intellectual and cultural milieu he was born into (Kenzer 1985a), given the historical context of late nineteenth-century German immigration into Warren County, Missouri, and considering the academic bent of his mother and father, the odds were certainly in Carl Sauer's favor that he would be of a scholarly leaning, and that his world view would be in accordance with that of his contemporaries. Indeed, everything about Sauer's childhood suggests the formation of a *Weltanschauung* similar to if not synonymous with that of the Warrenton community of which he was a part. Several years hence, his encounter with the C.W.C. *Weltbild* would reinforce and cement his early world view into place. But it was his father's influence—both at home and at college—which may actually have been the most important factor in the formation of his early world view.

William Albert Sauer: A Profile

Born in Beinstein (Remstal), Germany and always fond of his homeland, William Albert Sauer was perhaps the archetypical German academic of his era (Fig. 2). On the anniversary of his twenty-fifth year of service to Central Wesleyan College, the school newspaper ran a short tribute to the esteemed professor. "If we were to describe him to our readers," the author(s) wrote, "we would first of all mention his upright German character . . ." (*College Star* 1898). According to a colleague who payed homage to the elder Sauer immediately following his death, (William) Albert was described as "the best that the religious and educational life of the Germany of other days produced . . ." (*Central Wesleyan Star* 1918). Of all the professors at C.W.C., W. A. Sauer was perhaps the most self-consciously German, and he encapsulated best the spirit and motivation behind the small Missouri college.

The Sauers apparently come from a long tradition of teachers and musicians. W. Albert Sauer "was educated in the public school and later graduated from a teachers' college [in Germany]" *Warrenton Banner* 1918). As a student, his early academic interests appear to have centered around music and languages (Pulse 1906), but music soon became his foremost concern. The elder Sauer was an accomplished composer and musician, and the head of Central Wesleyan's Music Department for many years. His poems and lyrics frequently embellished the pages of the college newspaper. One of his students, Louis Weber (1851-1931), went on to become one of Missouri's leading late nineteenth-century composers (Baldridge 1970, especially pp. 17-19).

Music was an important part of the Central Wesleyan experience and Sauer's department was considered as fine a music center as one was apt to

(William) Albert Sauer (1844-1918), Carl Sauer's father. W. A. Sauer was decidedly the most influential person in his son's early intellectual development.

find in the region. Shortly before the turn of the century, the C.W.C. Conservatory of Music was described as one that would "compare favorably with any in the West" (*College Star* 1893). The respect accorded this small department was due, in no small part, to William Albert Sauer's background and to his view of higher education. Trained and educated wholly in his native Germany, he was, in the words of a contemporary, "a thoroughgoing educator of the German type, [one] who abhors all sham and superficialness" (Pulse 1906). For W. A. Sauer, music was never considered an end in itself, nor was it a substitute for a proper, scholarly education; music and book learning went hand in hand:

> Students who pursue a music course very often overlook the necessity of a higher general education. A real musician, just like the master of any other branch of art, must always be a person of general culture, not a mere mechanic. He who aims at a situation in a college or any higher school, must not forget that no school will appoint a music teacher who possesses no other accomplishments but his music (*College Star* 1887).

51

Similarly, in 1890, the elder Sauer would recommend that all college students get a well-rounded education. If you are planning to become a specialist, he would write, "learn all about the world." Even if you intend to become a doctor, he adds, be sure to study grammar, geography, and math as well, because no doctor should know only about medicine or human health (Sauer 1890).

Professor Sauer believed that education, like true musical understanding, cannot be taught: it must be desired. It may be instructive to quote, at some length, from some words of advice he offered in 1891. While the passage is directed at prospective musicians, I believe it is indicative of the gentleman's attitude toward learning in general, regardless of the specific discipline or endeavor:

> It is a mistake so often made by music scholars to expect their musical education exclusively from the teacher. . . . In order to become a musician, one must live and breathe in a musical atmosphere. If your surroundings are not of that nature, you yourself must create such an atmosphere around you. Do not perform trashy music, because your friends cannot understand any better, but try to elevate not only your own, but also their taste. Use every opportunity of hearing good music; be slow in criticism; do not reject what, at first, you do not understand; remember musical taste and understanding must be developed just as any other mental power (Sauer 1891).

Just as his son would later write that, to know the inhabitants of an area, you must "become one of the people; live with them if possible; take part in their activities" (Jones and Sauer,[8] p. 521), and that we must immerse ourselves in a region to understand its culture and "personality" (e.g., Sauer 1941a, 1941b), William Albert Sauer believed that you must likewise devote yourself, without reservation, to whatever enterprise you choose to undertake (also see Sauer 1888, Sauer 1893). In the same manner, we note that the elder Sauer refused to relinquish what he considered important; he would never settle for second best. It was with this same idealistic fervor that Carl Sauer refused to accept a second-rate geography; like his father, he too believed in quality above all else (see Kenzer 1985b, Stanislawski 1975).

Albert Sauer's Germanic heritage remained with him always, and his view of American culture can only be understood against this comparative backdrop. From one account we learn that "he is against all American superficiality concerning either politics or the church" (*College Star* 1898). In another instance the elder Sauer himself wrote that there is far too much individualism in this country, too much individual power. Alarmed that Americans possessed a false sense of "freedom," he expressed concern about America's preoccupation with the individual; America's progressive, secular, pragmatic culture and its associated lack of historical perspective worried him. Thus he contended that there is a distinct lack of respect for authority in the United States—no respect for adults, no regard for tradition, no sense of history (Sauer 1892a). By contrast, he believed that the Europeans had a greater sense of history and, as a consequence, a much stronger tie with their past. In his opinion, Europeans were therefore actually freer than Americans. A view to the past was extremely beneficial in his view and should *never* be compromised. Americans may be politically free, he would note, but he believed Europeans to be more reverent and more spiritually free due to their retrospective nature (see Sauer 1889).[9] It is not very difficult to see why his son was continually comparing American and German geography (Kenzer 1985b). This comparative practice was inherited from a father who straddled two distinct cultures.

In sum, America's progressive credo was antithetical to Professor Sauer's world view which, in turn, was synonymous with a common nineteenth-century German preoccupation with process and an historical perspective (see for example, Hughes 1958, pp. 183-248; Iggers 1983, pp. 3-173; Mead 1936, pp. 127-152). To the nineteenth-century German mind, the past is an ongoing phenomenon that fully explains the contemporary scene. One of W. A. Sauer's colleagues may have summarized this view in a talk when he spoke on "the value of historical study." "[A] knowledge of history," he argued, "brings the past up to the present, and how we thus find *the controlling powers,* and seeing *the causes* we find how the past is likened to the present . . ." (emphases added) (*College Star* 1899). As a product of nineteenth-century Germany, it was all but impossible for the elder Sauer (and his associates) not to think in historical terms. History and things historical were second nature to this generation of intellectuals. Historicism was more than a method. It was "an intellectual and scholarly movement which dominated historical, social, and humanistic studies in nineteenth-century Germany . . ." (Iggers 1973, p. 458; see Speth, this volume). As Hughes reminds us, even Max Weber, who was certainly no historian, was influenced by this same historical perspective:

> . . . the enormous merit of [late nineteenth-century] German social thought was that it dwelt *in the historical world.* History was one subject that Weber never specifically studied or taught. But his whole intellectual life was suffused with historical thinking. Law, like economics, was taught in Germany as a historical discipline. Sociology was being cast in a similar mold. And philosophy . . . had posed as one of its central problems the elaboration of the categories of historical thought (Hughes 1958, p. 293, emphasis in original).

It is important here to emphasize not only William Albert Sauer's German character and his preoccupation with the past, but to make note of his background in the physical and natural sciences (see Kenzer 1985a; also see *Warrenton Banner* 1918), in particular his interest in plants and geography. In addition to his duties as professor of music and French, he was also the college botanist. As well, he had a curiously persistent interest in maps. The study of geography was as important to him as a knowledge of history, and the combination of the two even more important. Fortunately, like his son, he was a rather opinionated individual who freely expressed and published his views on a variety of topics and issues. The following passage should prove insightful to those readers who may have wondered (a) where and when his son, Carl Sauer, first considered the value of historical geography, or (b) what the younger Sauer may have read as a child in Warrenton. In 1892, suggesting which German books a student should read, Professor Sauer wrote:

> Well known are the excellent historical works by Becker, Dittmer [*sic*], Ranke, Schlosser [*sic*] and Weber but I would like to recommend to my young friends, for their special purposes in particular, Redenbacher's primer of world history, which surely will remain a lifelong favorite handbook for them. Whoever reads about history should not omit to look up the area on the map where the event took place. A historical atlas, the quite reasonable one by Putzge [*sic*] will be of great value to him. And since we now have mentioned the relationship between history and geography, I would suggest for the purpose of self-instruction the primer of geography by Schwartz. If, however, someone would prefer a more scientific approach, I would suggest Daniel's small textbook on geography (Sauer 1892b).[10]

In his view, history and geography went hand in hand, one useless without the other, both a necessary component of meaningful research. Even in a discussion on literary works, he would suggest that "when reading such works of literature, [be sure] to keep the textbook of history plus the map nearby in order to place everything in chronological and geographical order" (Sauer 1907, p. 9).

One thing is surely evident: Carl Sauer's early recognition of the relationship between history and geography (see Sauer 1916, pp. 144f.; Sauer 1918, pp. 45-83; Sauer 1920b, pp. 73-174; Sauer 1924, pp. 18-19; Sauer 1925) and his subsequent dicta on the subject (Sauer 1941a, 1952, 1984) did not arise out of the blue. Whether he was aware of it or not, he was mirroring an attitude that was traditionally German on the one hand (see Pfeifer 1965, James 1968, p. 9) and specific to his father's own world view on the other. Like his father, he was very conscious of the influence of the past. Even when the younger Sauer wrote a paper on the *current* economic problem of his Ozark homeland, he qualified his approach, arguing that "for an understanding of the area it is essential to keep in mind its antecedents, and also that the blood of the frontiersman is still dominant among the population" (Sauer 1920a, p. 217).

Goethean History

The C.W.C. intellectual climate was heavily suffused with a conception of history based on the writings of Johann Wolfgang von Goethe (1749-1832). Indeed, the institution's *Weltbild,* the professors' respective philosophies and, consequently, the students themselves, were all fully enveloped in this distinctive intellectual environment. Goethe's maxims filled the pages of the school's newspaper and journal. The oldest student organization on campus was the Goethenia Society—a literary association—and Carl Sauer became a devoted member as an undergraduate (Kenzer 1985a, pp. 263-264). The society took its name and motto (*"Mehr Licht"*) from Goethe, and there is no way to understand either the C.W.C. experience or Sauer's undergraduate heritage outside of this context.

While much of nineteenth-century German historicism was grounded in an analysis and explanation of political history (i.e., history of the state) —what Iggers calls "the German historicist tradition" (Iggers 1983, p. 13)—the Goethean strain of German humanistic-historical thought viewed culture (and the individual) as the focal point of history (Holborn 1970). This was the predominant German social philosophy of the nineteenth century, a "German national philosophy" adumbrated in the late eighteenth century and given full substance in the writings of Fichte, Hegel, and Schelling—the "great idealists" of the early nineteenth century (Beck 1967, pp. 301-307). This was likewise the very basis of Goethe's morphology—i.e., the understanding of form as expressed creatively (culturally/individually). Form—the manifestation of human, spiritual, and natural processes—was *the* single most important element in Goethe's world view. "Consequently," notes Bergstraesser, "he saw history as essentially cultural history. His interest in it was centered upon the forms by which man lives and conceives his own existence and by which he gives expression to this experience in religion, art, and science" (1962, pp. 205-206).

Speth (1981) and Williams (1983) have demonstrated that Carl Sauer was writing in a Goethean intellectual tradition. Sauer not only acknowledges this debt to Goethe—both in his published works and in his correspondence— but uses Goethe's very term ("morphology") to express the essence of his important essay on geographic methodology (Sauer 1925). Sauer's form-laden, phenomenological outline for a cultural, experiential geography was predicated on a Goethean conception of humanity and thus, like Goethe, Sauer tried to reconstruct the inner meaning and the forms of historical cultures (see Bergstraesser 1962, p. 209). By defining geography as culture history, Sauer was merely attempting "to reestablish the German classical geographical tradition in American geography" (Entrikin 1984, p. 405). *But Sauer's cultural-historical geography was based not on a reading of Goethe, but on the fact that Sauer had been totally immersed in a Goethean world view since birth.* This was further compounded by his three-year stay in southern Germany as a child (1898-1901) and later, perhaps more so, by his association with the C.W.C. intellectual environment and his enduring involvement with the Goethenia Society as an undergraduate (Kenzer 1985a).

Sauer was certainly no stranger to Goethe's writings, nor to the intellectual legacy Goethe left to nineteenth-century Germany. In his attempt to place American geography on firmer ground than it enjoyed in the 1920s, Sauer simply had to reach into his past and build on a familiar Germanic heritage. It was a Goethean heritage, one that stressed morphology and personal, real world experience; *not* an Hegelian heritage that sought to explain via universal laws. Indeed, Goethe's inductive, experiential approach is what best characterizes his writings from those of his contemporaries. As Stern so correctly writes, ". . . for Goethe, the aesthetic is a full subsuming, not an aestheticist attenuation, of experience" (1971, p. 130).

Goethe was a complex individual and his writings show a man caught between two intellectual worlds. His early life and ideas clearly reflect the climate of romanticist, late eighteenth-century Germany. By the end of Goethe's career, however, German intellectual thought began to shift away from the classical romanticism of his youth, and was instead becoming a "science" immersed in idealism and the experimental philosophy of positivism (see Speth, this volume). A close analysis of Goethe's works reveals that "the sage of Weimar" decidedly straddled both worlds (see Nisbet 1972). In retrospect, his writings consequently seem ambiguous or inconsistent, as they indeed were:

> He picked out an idea here and there which he came across in rather haphazard reading . . . and he did not trouble himself about the logical proof of the idea which interested him, nor about the steps which followed from it, but took these thoughts because they suited him and worked them into the complex weave of his own emotions and beliefs (Trevelyan 1949, p. 124).

This lack of logic is perhaps a fitting tribute to a man who is remembered mainly for his poetry and novels, even though Goethe considered himself a scientist and felt that his literature could not be understood without first enquiring into his science (see Zweig 1967). It is of little surprise that he has been called an illogical philosopher (Trevelyan 1949, p. 122). As a philosopher of objects, a philosopher of spirit and material forms, a philosopher of experience and metamorphosis, he was not a philosopher of the traditional type: "I lacked the organ for philosophy in the proper sense" (quoted in Magnus 1949, p. 235; also see Bréhier 1968, pp. 220-221). Like his intellectual descendant Heinrich Heine (1797-1856), Goethe was leery of many of his contemporaries and "both [Goethe and Heine] were critical, sometimes scathingly so, of the overwhelmingly intellectual and philosophically informed culture of their ages" (Stern 1971, p. 55). Goethe's philosophy was admittedly inconsistent.

By the same token, a follower of Goethe—whether it be an institution (C.W.C.) or an individual (Carl Sauer)—cannot be expected to be entirely logical or consistent either. (Moreover, logic and consistency are value-laden concepts, and both must be judged relative to prevailing epistemological frameworks.) Thus it becomes as no great surprise to find Sauer characterized as paradoxical (Hooson 1981, p. 166), intransigent (West 1979, p. 35), or as an "intellectual Voortrekker" (Williams 1983, p. 2). Like Goethe, Sauer would pick and choose ideas and concepts to fit the occasion. As Entrikin has recently observed, for Sauer an hypothesis "was a means to an end" and nothing more (1984, p. 390). In Sauer's eyes, an argument was merely "a means of solving logical puzzles that arose in the course of specific empirical studies . . ." (Entrikin 1984, p. 387). "Sauer's goal," he points out, "was to work through conceptual problems that he encountered in his empirical, field-oriented studies, not to establish a consistent logical framework or system" (Entrikin 1984, p. 387). This equates with Goethe's maxim that "truth is individual and, although it is such, or rather because it is such, is true" (Croce 1923, p. 15). Absolute truth and consistency were consequently as unimportant to Sauer as they were to Goethe.

A Short Conclusion

In a recent paper it was postulated that a partial misunderstanding of Sauer's complete world view has resulted, due, in part, to a dearth of research into his pre-Berkeley intellectual inheritance (Entrikin 1984, p. 389). This paper is thus a response to that premise and a second step in identifying the contexts of Sauer's earliest experiences and intellectual development (Kenzer 1985a). Like the rest of us, Sauer was born into a particular social, cultural, and intellectual context—a certain milieu where recognizable associations of people and events acting in space and through time can be identified—and he consciously and unconsciously came to accept that context as "reality." It was this first milieu, his initial "construction of reality" (Berger and Luckmann 1980), that I am arguing was the most crucial for an understanding of Sauer's ideas and publishing writings. Essentially, I subscribe to the view expressed by Anderson:

> Men and movements are not born in a vacuum; neither do they find their fulfillment apart from temporal and social considerations. A productive soil is just as necessary for the full development of ideas and institutions as it is for botanical species (1940, p. 478).

The aim, then, has been to expose the "soil" into which Sauer was born, within which he developed his first world view. Thenceforth, his *Weltanschauung* acted as a sort of filter by which he screened his experiences and through which everything he read and assimilated had to pass.

Sauer's anthropogeographic approach to the study of geography, his reliance on an historical mode of explanation, his emphasis on culture history, and his insistence on a scholarly (though inductive) orientation, are all elements taken from turn-of-the century German social science, more specifically, German geographic thought (see Leighly 1938; cf. Hartshorne 1939, pp. 260-277 [84-101]. The purpose of this paper has been, firstly, to indicate the degree to which Sauer was steeped in a strong Germanic world view from birth and, secondly, to note how he adopted that world view through his peers and family in Warrenton. I do not claim that Sauer was totally conscious of this fact, but there seems to be little doubt, given the social, cultural, and intellectual climate of his hometown and undergraduate institution, that he had much of a say-so in the matter. More importantly, however, it has been my intent to show that an informed interpretation of Sauer's geography and published writings *must* take into account his deep Goethean heritage and Germanic upbringing. In essence, I am suggesting that Carl Sauer was perhaps doing no more than perpetuating the Central Wesleyan College tradition while following in the footsteps of his father.

Acknowledgments

I would like to express thanks to Robert A. Rundstrom and William W. Speth for their comments on an earlier draft of this paper. I would also like to thank Carol Fowler-Dage and Odessa Ofstad for their respective contributions in accessing the Central Wesleyan College Archives at Northeast Missouri State University, and for Ms. Ofstad's permission to reproduce Figure 1. As well, I wish to thank Elizabeth Sauer FitzSimmons for allowing me to examine her father's private correspondence.

Notes

1. I rely a good deal on Mack Walker's masterly study for this particular section (Walker 1964). As a companion to Walker, specifically with regard to emigration from Westphalia to Missouri, see Kamphoefner 1978 and Kamphoefner 1982. For an introduction to American immigration the following sources should be of some help: Hansen 1940, Jones 1960, Taylor 1971, Archdeacon 1983.

2. The German Revolution of 1848 was, of course, only one of numerous European social uprisings which occurred between 1848 and 1849. This, in part, accounts for the large number of individuals who chose to emigrate to the New World, rather than to other parts of Europe during this generally unsettled period. It also suggests a possible explanation for the small number of those who actually returned to Germany after the nation achieved a degree of stability. The entire European political environment was fragile and many feared that additional hostilities might arise. For an introduction to the overall European situation at the time, I find the following sources helpful: Fejtö 1948, Robertson 1952, Maurice 1969, Stearns 1974.

3. Johnson has indicated that Germans were more evenly distributed across America than all other "foreign born groups" by 1870 (1951, p. 3).

4. What follows is a complete list of C.W.C. professors and teaching assistants for the academic year 1907-08: G. B. Addicks, J. H. Frick, W. A. Sauer, H. Vosholl, J. M. Rinkel, C. J. Stueckman, O. E. Kriege, C. L. Wellemeyer, A. W. Ebeling, E. Weiffenbach, G. C. Hohn, Z. Nagel, M. L. Nagel, M. M. Drew, E. Haenssler, I. Hartel, E. Zimmermann, A. Schulze, C. Werner, C. Heidel, L. Nagel, P. Walter, C. Sauer, P. Ditzen, J. T. Myers, F. O. Kettelkamp, I. Schoeppel, C. Bader, G. V. Tungeln, and E. H. Kassmann (Annual Catalogue 1908, pp. 6-7).

5. The "St. Louis School" is generally attributed to William T. Harris (1835-1909) who began *The Journal of Speculative Philosophy* based on the idealistic writings of Hegel, particularly his *Logic*. Like Hegel, Harris argued that science was grounded in experience and to comprehend reality— indeed, to understand existence at all—we must realize and accept the dialectical nature of the universe: in sum, the notion that everything is in a continual state of reciprocal change and thus the only two constants are time and space (and their mutual interaction). Harris's ideas became very popular in the St. Louis area and notably among the local intellectuals who formed the St. Louis Society. For an introduction to Harris and his school see the following sources: Forbes 1930,

especially pp. 83-90; Forbes 1931; Runes 1955, pp. 466-469; Easton 1967. A good starting point for information on Quincy, Illinois, especially with respect to Hegelian thought and the general philosophical climate of the area at the time, is Anderson (1940, 1941).

6. John Leighly, Carl Sauer's friend and colleague for over fifty years, never mentioned the existence of Sauer's older brother in either of the two lengthy reminiscences he wrote following Sauer's death in 1975 (Leighly 1976, 1978). In a subsequent letter to the author, Leighly has acknowledged that he never even knew there was an older brother until after Carl Sauer had died (personal correspondence, March 23, 1983)!

7. For an interesting account of some of the differences between German-born women and their contemporary German-American counterparts, see Billigmeier 1974, pp. 66-68; cf. Faust 1969, vol. 2, pp. 448-464. Also see the more recent article on stereotypes of German-American women in Missouri by Pickle (1985).

8. It is often assumed, because of the order of their names, that Jones was the senior and Sauer the junior author of this influential essay. Discussing the article twenty-four years after its publication, however, the two men reveal that this arrangement was merely alphabetical and did not imply any senior-junior designation (*W. D. Jones to C. O. Sauer, December 22, 1939*, Sauer Papers; see Kenzer 1985b).

9. It would appear that the elder Sauer's strong religious training prevented him from separating politics from religion as his son was able to do (see Entrikin 1984), though Carl Sauer's rejection of his religious heritage was a late occurrence (Kenzer 1985a) and something that seems to have come about gradually.

10. The German writers to whom Professor Sauer was referring were probably Wilhelm Adolf Becker (1796-1846), Heinrich Dittmar (1792-1866), Leopold von Ranke (1795-1886), August Ludwig von Schlözer (1735-1809), Max Weber (?) (1864-1920), Wilhelm Redenbacher (1800-1876), Friedrich Wilhelm Putzger (b. 1849), Friedrich Leberecht Wilhelm Schwartz (1821-1899), and Hermann Adalbert Daniel (1812-1871). Few of these books were part of the Central Wesleyan library, at least not in 1877 (Books in the Library 1877). It is therefore likely that Sauer was looking through his personal library when naming these important history and geography texts.

Literature Cited

Anderson, P. R. "Hiram K. Jones and Philosophy in Jacksonville." *Journal,* Illinois State Historical Society, vol. 33, no. 4 (December 1940): 478-520.

Anderson, P. R. "Quincy, an Outpost of Philosophy." *Journal,* Illinois State Historical Society, vol. 34, no. 1 (March 1941): 50-83.

Annual Catalogue. *Forty-Fourth Annual Catalogue, Central Wesleyan College, 1907-1908.* Warrenton, Missouri: Central Wesleyan College, 1908.

Archdeacon, T. J. *Becoming American: An Ethnic History.* New York: The Free Press, 1983.

Baldridge, T. L. "Louis Weber (1851-1931), Kansas City Composer and Publisher." Unpublished Master of Music in Musicology Degree, University of Kansas, 1970.

Beinsteiner Heimatblatt. "Aus Beinstein." No. 12 (December 1920): 4.

Beck, L. W. "German Philosophy." In *The Encyclopedia of Philosophy,* edited by P. Edwards, pp. 291-309. Vol. 3. New York: Macmillan Publishing Co. and The Free Press, 1967.

Bek, W. G. *The German Settlement Society of Philadelphia and its Colony Hermann, Missouri.* Philadelphia: Americana Germanica Press, 1907.

Berger, P. L., and T. Luckmann. *The Social Construction of Reality: A Treatise in the Sociology of Knowledge.* New York: Irvington Publishers, 1980.

Bergstraesser, A. *Goethe's Image of Man and Society.* Freiburg: Herder, 1962.

Billigmeier, R. H. *Americans From Germany: A Study in Cultural Diversity.* Belmont, California: Wadsworth Publishing Company, 1974.

Blackbourn, D. "The Discreet Charm of the Bourgeoisie: Reappraising German History in the Nineteenth Century." In *The Peculiarities of German History: Bourgeois Society and Politics in Nineteenth-Century Germany,* edited by D. Blackbourn and G. Eley, pp. 157-292. Oxford and New York: Oxford University Press, 1984.

Books in the Library. *Catalogue of Books in the Library of the Central Wesleyan College, August 1st, 1877.* Central Wesleyan College Archives: 2.3/1.

Bratton, S. T., and M. Langendoerfer. "The Hermann, Missouri Region." *Bulletin,* Geographical Society of Philadelphia, vol. 29, no. 2, (1931): 115-129.

Bréhier, E. *The Nineteenth Century: Period of Systems, 1800-1850.* Translated by Wade Baskin. Chicago and London: University of Chicago Press, 1968.

Central Wesleyan Star. ["In Memoriam."] Vol. 36, no. 1 (October 1918): 1.

College Star. ["Our Music Department."] Vol. 10, no. 6 (March 1893): 3.

College Star. ["Prof. W. A. Sauer."] Vol. 15, no. 8 (May 1898): 3.

College Star. ["Professor Sauer on a Musician's Education."] Vol. 5, no. 2 (November 1887): 3.

College Star. ["The Value of Historical Study."] Vol. 17, no. 2 (November 1899): 7.

Collier, J. E. "Geography of the Northern Ozark Border Region in Missouri." *The University of Missouri Studies* 26. Columbia: University of Missouri, 1953.

Croce, B. *Goethe.* Translated by Emily Anderson. London: Methuen & Co., 1923.

Duden, G. *Bericht über eine Reise nach den westlichen Staaten Nordamerika's und eined mehrjährigen Aufenthalt am Missouri.* Elberfeld: Sam Lucas, 1829.

Easton, L. D. "Harris, William Torrey." In *The Encyclopedia of Philosophy,* edited by P. Edwards, pp. 416-417. Vol. 3. New York: The Macmillan Company and The Free Press, 1967.

Ellis, J. F. *The Influence of Environment on the Settlement of Missouri.* St. Louis: Webster Publishing Company, 1929.

Entrikin, J. N. "Carl O. Sauer, Philosopher in Spite of Himself." *Geographical Review,* vol. 74, no. 4 (October 1984): 387-408.

Faust, A. B. *The German Element in the United States.* Two volumes. New York: Arno Press and the New York Times, 1969.

Fejtö, F., editor. *The Opening of an Era, 1848: An Historical Symposium.* London: Allan Wingate, 1948.

FitzSimmons, Elizabeth Sauer to Martin S. Kenzer, personal correspondence, October 16 and October 27, 1984.

Forbes, C. "The St. Louis School of Thought." *Missouri Historical Review,* vol. 25, no. 1 (October 1930): 83-101.

Forbes, C. "The St. Louis School of Thought." *Missouri Historical Review,* vol. 26, no. 1 (October 1931): 68-77.

Gerlach, R. L. "German Settlements in the Ozarks of Missouri." *Rundschau: American-German Review,* vol. 3 (May 1973): 6-7.

Gerlach, R. L. *Immigrants in the Ozarks: A Study in Ethnic Geography.* Columbia and London: University of Missouri Press, 1976a.

Gerlach, R. L. "Population Origins in Rural Missouri." *Missouri Historical Review,* vol. 71, no. 1 (October 1976b): 1-21.

Hansen, M. L. *The Atlantic Migration, 1607-1860: A History of the Continuing Settlement of the United States.* Cambridge: Harvard University Press, 1940.

Hartshorne, R. "The Nature of Geography: A Critical Survey of Current Thought in the Light of the Past." *Annals,* Association of American Geographers, vol. 29, nos. 3 and 4 (September and December 1939): 173-412 [i-236] and 413-658 [237-482].

Haselmayer, L. A. "Das Deutsche Kollegium: Wesleyan's Teutonic Past." *Annals of Iowa,* 3rd series, vol. 35, no. 3 (Winter 1960): 206-215.

Haselmayer, L. A. "German Methodist Colleges in the West." *Methodist History,* n.s., vol. 2, no. 3 (July 1964): 35-43.

Hawgood, J. A. *The Tragedy of German-America.* New York: Arno Press and the New York Times, 1970.

Herbst, J. *The German Historical School in American Scholarship: A Study in the Transfer of Culture.* Ithaca, New York: Cornell University Press, 1965.

Hinsdale, B. A. "Notes on the History of Foreign Influence Upon Education in the United States." *United States, Bureau of Education, Report of the Commissioner for 1897/98.* Vol. 1. Washington, D.C. 1899.

Holborn, H. "German Idealism in the Light of Social History." Translated by Robert Edwin Herzstein. In *Germany and Europe: Historical Essays by Hajo Holborn,* pp. 1-32. Garden City, New York: Doubleday & Company, 1970.

Hooson, D. "Carl O. Sauer." In *The Origins of Academic Geography in the United States,* edited by B. W. Blouet, pp. 165-174. Hamden, Connecticut: Archon Books, 1981.

Hughes, H. S. *Consciousness and Society: The Reorientation of European Social Thought, 1890-1930.* New York: Alfred A. Knopf, 1958.

Iggers, G. G. "Historicism." In *Dictionary of the History of Ideas: Studies of Selected Pivotal Ideas,* edited by P. P. Wiener, pp. 456-464. Vol. 2. New York: Charles Scribner's Sons, 1973.

Iggers, G. G. *The German Conception of History: The National Tradition of Historical Thought from Herder to the Present.* Second edition. Middletown, Connecticut: Wesleyan University Press, 1983.

James, P. E. "Continuity and Change in American Geographic Thought." In *Geography and the American Environment,* edited by S. B. Cohen, pp. 2-14. Washington: Voice of America Forum Lectures, 1968.

Johnson, H. B. "The Location of German Immigrants in the Middle West." *Annals,* Association of American Geographers, vol. 41, no. 1 (March 1951): 1-41.

Jones, M. A. *American Immigration.* Chicago: University of Chicago Press, 1960.

Jones, W. D., and C. O. Sauer. "Outline for Field Work in Geography." *Bulletin,* American Geographical Society, vol. 47, no. 7 (1915): 520-525.

Kamphoefner, W. D. "Transplanted Westfalians: Persistence and Transformation of Socioeconomic and Cultural Patterns in the Northwest German Migration to Missouri." Unpublished Ph.D. Dissertation, University of Missouri, Columbia, 1978.

Kamphoefner, W. D. *Westfalen in der Neuen Welt: Eine Sozialgeschichte der Auswanderung im 19. Jahrhundert.* Münster: F. Coppenrath Verlag, 1982.

Kargau, E. D. "Missouri's German Immigration." *Missouri Historical Society Collections,* vol. 2, no. 1 (January 1900): 23-34.

Kenzer, M. S. "Milieu and the 'Intellectual Landscape': Carl O. Sauer's Undergraduate Heritage." *Annals,* Association of American Geographers, vol. 75, no. 2 (June 1985a): 258-270.

Kenzer, M. S. "Carl Sauer and the Carl Ortwin Sauer Papers." *History of Geography Newsletter,* no. 5 (December 1985b). In press.

Leighly, J. "Methodologic Controversy in Nineteenth Century German Geography." *Annals,* Association of American Geographers, vol. 28, no. 4 (December 1938): 238-258.

Leighly, J. "Carl Ortwin Sauer, 1889-1975." *Annals,* Association of American Geographers, vol. 66, no. 3 (September 1976): 337-348.

Leighly, J. "Carl Ortwin Sauer: 1889-1975." In *Geographers: Biobibliographical Studies,* edited by T. W. Freeman and P. Pinchemel, pp. 99-108. Vol. 2. London: Mansell, 1978.

Magnus, R. *Goethe as a Scientist.* Translated by Heinz Norden. New York: Henry Schuman, 1949.

Mandelbaum, M. *History, Man, and Reason: A Study in Nineteenth Century Thought.* Baltimore and London: The Johns Hopkins University Press, 1971.

Maurice, C. E. *The Revolutionary Movement of 1848-9 in Italy, Austria-Hungary, and Germany.* New York: Greenwood Press, 1969.

McClelland, C. E. *State, Society, and University in Germany, 1700-1914.* Cambridge, England: Cambridge University Press, 1980.

Mead, G. H. *Movements of Thought in the Nineteenth Century.* Chicago: University of Chicago Press, 1936.

Mendelsohn, E. "The Biological Sciences in the Nineteenth Century: Some Problems and Sources." *History of Science,* vol. 3 (1964): 39-59.

Namier, L. *1848: The Revolution of the Intellectuals.* The Raleigh Lecture on History, 1944, from the Proceedings of the British Academy, vol. 30. London: Oxford University Press, 1944.

Nisbet, H. B. *Goethe and the Scientific Tradition.* London: University of London Institute of Germanic Studies, 1972.

O'Boyle, L. "Learning for its Own Sake: The German University as Nineteenth-Century Model." *Comparative Studies in Society and History,* vol. 25, no. 1 (January 1983): 3-25.

O'Connor, R. *The German-Americans: An Informal History.* Boston and Toronto: Little, Brown and Company, 1968.

Paul, R. "German Academic Science and the Mandarin Ethos, 1850-1880." *British Journal for the History of Science,* vol. 17, part 1, no. 55 (March 1984): 1-29.

Pfeifer, G. "Carl Ortwin Sauer zum 75. Geburtstage am 24. XII. 1964." *Geographische Zeitschrift,* vol. 53 (February 1965): 1-9.

Pickle, L. S. "Stereotypes and Reality: Nineteenth-Century German Women in Missouri." *Missouri Historical Review,* vol. 79, no. 3 (April 1985): 291-312.

Pulse. *'06 Pulse [C. W. C.].* Warrenton, Missouri: Central Wesleyan College, 1906.

Rafferty, M. D. *Historical Atlas of Missouri.* Norman: University of Oklahoma Press, 1982.

Ringer, F. K. *The Decline of the German Mandarins: The German Academic Community, 1890-1933.* Cambridge: Harvard University Press, 1969.

Rippley, L. J. *The German-Americans.* Boston: Twayne Publishers, 1976.

Robertson, P. *Revolutions of 1848: A Social History.* Princeton: Princeton University Press, 1952.

Roueché, B. *Special Places: In Search of Small Town America.* Boston and Toronto: Little, Brown and Company, 1982.

Runes, D. D. *Treasury of Philosophy.* New York: Philosophical Library, 1955.

Sauer, C. O. *Geography of the Upper Illinios Valley and History of Development.* Bulletin No. 27. Illinois Geological Survey. Urbana: University of Illinois, 1916.

Sauer, C. O. "Geography." In *Starved Rock State Park and Its Environs,* by C. O. Sauer, G. H. Cady, and H. C. Cowles, pp. 3-83. Bulletin No. 6. Geographic Society of Chicago. Chicago: University of Chicago Press, 1918.

Sauer, C. O. "The Economic Problem of the Ozark Highland." *Scientific Monthly,* vol. 11 (September 1920a): 215-227.

Sauer, C. O. *The Geography of the Ozark Highland of Missouri.* Bulletin No. 7. The Geographical Society of Chicago. Chicago: University of Chicago Press, 1920b.

Sauer, C. O. "The Survey Method in Geography and its Objectives." *Annals,* Association of American Geographers, vol. 14, no. 1 (March 1924): 17-33.

Sauer, C. O. "The Morphology of Landscape." *University of California Publications in Geography,* vol. 2, no. 2 (October 1925): 19-54.

Sauer, C. O. "Foreword to Historical Geography." *Annals,* Association of American Geographers, vol. 31, no. 1 (March 1941a): 1-24.

Sauer, C. O. "The Personality of Mexico." *Geographical Review,* vol. 31, no. 3 (July 1941b): 353-364.

Sauer, C. O. "Folkways of Social Science." In *The Social Sciences at Mid-Century: Papers Delivered at the Dedication of Ford Hall, April 19-21, 1951,* pp.100-109. Minneapolis: University of Minnesota Press, 1952.

Sauer, C. O. "Regional Reality in Economy." Edited and with commentary by Martin S. Kenzer. *Yearbook,* Association of Pacific Coast Geographers, vol. 46 (1984): 35-49.

Sauer, W. A. "Worin es manche Studenten verfehlen." *College Star,* vol. 6, no. 1 (October 1888): 7-8.

Sauer, W. A. "Ehrfurcht—und warum es daran fehlt." *College Star,* vol. 6, no. 7 (April 1889): 6-7.

Sauer, W. A. "Missgriffe junger Leute." *College Star,* vol. 8, no. 1 (October 1890): 5.

Sauer, W. A. "Some Advice to Music Scholars." *College Star,* vol. 9, no. 3 (December 1891): 6.

Sauer, W. A. "Einige gedanken über 'Freiheit'." *College Star,* vol. 10, no. 2 (November 1892a): 7-8.

Sauer, W. A. "Was für deutsche Bücher sollen wir lesen?" *College Star,* vol. 10, no. 3 (December 1892b): 6-7.

Sauer, W. A. "System in Study." *College Star,* vol. 11, no. 1 (October 1893): 5-6.

Sauer, W. A. "Wo fehlt's? Eine Literarische Ermahuung." *Central Wesleyan Star,* vol. 24, no. 4 (January 1907): 9-10.

Schneider, C. E. *The German Church on the American Frontier: A Study in the Rise of Religion among the Germans of the West.* St. Louis: Eden Publishing House, 1939.

Schultz, G. *Early History of the Northern Ozarks.* Jefferson City, Missouri: Midland Printing Company, 1937.

Speth, W. W. "Berkeley Geography, 1923-33." In *The Origins of Academic Geography in the United States,* edited by B. W. Blouet, pp. 221-244. Hamden, Connecticut: Archon Books, 1981.

Stanislawski, D. "Carl Ortwin Sauer, 1889-1975." *Journal of Geography,* vol. 74, no. 9 (December 1975): 548-554.

Stearns, P. N. *1848: The Revolutionary Tide in Europe.* New York: W. W. Norton and Company, 1974.

Stern, J. P. *Idylls & Realities: Studies in Nineteenth-Century German Literature.* London: Methuen and Co., 1971.

Taylor, P. *The Distant Magnet: European Emigratioin to the U.S.A.* London: Eyre and Spottiswoode, 1971.

Trevelyan, H. "Goethe as Thinker." In *Essays on Goethe,* edited by W. Rose, pp. 122-140. London: Cassell and Co., 1949.

Turner, F. J. "German Immigration in the Colonial Period." *Chicago Record-Herald,* August 28, 1901:7.

Turner, F. J. *The Frontier in American History.* New York: Henry Holt and Company, 1920.

Viereck, L. *German Instruction in American Schools.* New York: Arno Press, 1978.

Walker, M. *Germany and the Emigration, 1816-1885.* Cambridge: Harvard University Press, 1964.

Warrenton Banner. ["Death of W. A. Sauer."] August 30, 1918: 1.

Warrenton Banner. ["Death of H. A. Sauer."] October 16, 1936a: 1.

Warrenton Banner. ["Death of H. A. Sauer."] October 23, 1936b: 4.

Warrenton Banner. ["Death of R. J. Sauer."] July 30, 1942: 1.

Warrenton Herald. ["Return of Sauer Family."] August 21, 1901: 4.

West, R. C. *Carl Sauer's Fieldwork in Latin America.* Ann Arbor: University Microfilms International, 1979.

Williams, M. "'The Apple of my Eye': Carl Sauer and Historical Geography." *Journal of Historical Geography,* vol. 9, no. 1 (January 1983): 1-28.

Wittke, C. *Refugees of Revolution: The German Forty-Eighters in America.* Philadelphia: University of Pennsylvania Press, 1952.

Wittke, C. *We Who Built America: The Saga of the Immigrant.* Revised edition. Cleveland: Case Western Reserve University Press, 1967.

Wittke, C. *The German-Language Press in America.* New York: Haskell House Publishers, 1973.

Wyman, M. *Immigrants in the Valley: Irish, Germans, and Americans in the Upper Mississippi Country, 1830-1860.* Chicago: Nelson-Hall, 1984.

Zweig, A. "Goethe, Johann Wolfgang von." In *The Encyclopedia of Philosophy,* edited by P. Edwards, pp. 362-364. Vol. 3. New York: The Macmillan Company and The Free Press; London: Collier-Macmillan, 1967.

PART TWO

Research Directions

Preparing for the National Stage: Carl Sauer's First Ten Years at Berkeley

Anne Macpherson

Carl Ortwin Sauer came to the University of California at Berkeley in 1923 to modernize the Department of Geography. He had favorably impressed Ruliff S. Holway (1857-1927), then Chairman of Geography at the University of California, as a man who could build a distinguished department. Sauer was attracted to California by the prospect of new fields of investigation, a new area in which to "practice his trade." During his first ten years at Berkeley, Sauer worked with a number of social scientists on projects that determined the direction of his life's work, had a great impact on the social sciences at the university, and brought him to national attention.

Geography and Social Science in the 1920s

In the 1920s, geography was a junior partner of geology in most institutions. Geography at Berkeley was an independent department, although it was housed in Bacon Hall with Geology. The "technical side" of geography, physiography, was dominated by William Morris Davis's (1850-1934) theory of the cycle of erosion. Climatology was in its infancy, Berkeley being one of the few departments where any work or courses were offered in the subject. "Commercial Geography" was the second strong area of the field. "Modern Geography," including the human element, investigated how the physical environment influenced human behavior, economy, or destiny.

With few exceptions, social studies were recent university subjects, and only nascent sciences, developing as such during the troubled times of the Great Depression, between the two world wars. Two institutions played a significant role in promoting the social sciences in the United States: the Social Science Research Council and the Laura S. Rockefeller Memorial Fund.

The Social Sciences Research Council (S.S.R.C.) was established in 1923 as a corporation with endowed funds, governed by a council of thirty who were also its only members. Twenty-one of the members were appointed from seven professional associations and nine were at large, one being Isaiah Bowman (1878-1950) of the American Geographical Society. The council's purpose was to improve research and teaching of social sciences in the United States, and it was empowered to raise funds for its projects. It worked through appointed committees, and could disburse funds for fellowships and grants-in-aid for individual research. It also could undertake projects to investigate problems of national importance or conditions affecting social science research. The S.S.R.C. was especially interested in projects that cut across traditional fields of knowledge, involving coordinated efforts. It advised and assisted the Rockefeller Foundation, and was partially funded by it.

The Laura S. Rockefeller Memorial Fund also was established in 1923, under the direction of Beardsley Ruml. The senior Rockefeller Foundation was chiefly interested in medicine. The policy of the Memorial Fund was to bring disciplines together to investigate social problems, and to promote long-term projects. Under Ruml, research was to be directed toward the improvement of human welfare. After the Memorial Fund was merged with the Rockefeller Foundation in 1929 the policy changed, allowing more research as an end in itself, without direct application to contemporary problems. Foundation aid was concentrated in high-grade university centers selected on a regional basis. In addition to the extensive use of fellowships, one of the principal techniques to encourage improved research was to support committees or councils at various universities that would control and administer the fluid research funds given by the foundation. Each institution could then itself determine the fields or projects to be supported (Fosdick 1972). The foundation maintained this policy until 1934 when growing dissatisfaction with the results and handling of the fluid funds forced a reexamination. After 1934 the foundation promoted projects with some hope of application to social problems. After 1937 the Rockefeller Foundation developed a new interest in Latin America, in conjunction with the American Council of Learned Societies (A.C.L.S.).

The A.C.L.S. was organized in committees with a regional or topical focus. The Association of American Geographers (A.A.G.) did not join until 1939, on which occasion Sauer wrote to Preston James at the University of Michigan:

> I don't know what's coming over the A. A. Geographers. It surely can't be that their interest in the American Council of Learned Societies is perking up because the A.C.L.S. might get some money for Latin American studies. I wouldn't suspect them of any motive as low as that, but rather, as you stated, that they "have an important role to play in the dissemination of geographical ideas." By all means let the missionary work go on. In years past when I have occasionally asked the A.A.G. to affiliate with the Social

Science Research Council, I have been given some variant of the position that geography was different from and perhaps greater than social science. In this case I take it there is no such difficulty, because the A.A.G. certainly would not declare that it is not a learned society (*Sauer to James, March 2, 1939,* SP[1]).

This letter was written on the same day that Sauer wrote Isaiah Bowman accepting nomination as President of the A.A.G. Sauer had long been recognized and honored nationally outside the field of geography.

Carl Sauer in Berkeley

In the 1920s Geography at Berkeley was a two-man department, with four teaching fellows who were not necessarily geography students. Ruliff S. Holway, a "modest man" according to John Leighly, was a physiographer, but aware of trends in the field, which he termed "modern geography." A member of the A.A.G., he had met Sauer at the annual meeting in 1920 and was impressed by him enough to write U. C. President David P. Barrows to nominate Sauer for a position in the department:

> The present staff of the Department in the University of California is insufficient and the schedule too one-sided to present properly the modern conception of Geography. There is a great need of a man to present the human element—the influence of Geographic conditions upon man and his activities. . . . We offered Dr. Sauer $2,000 in 1917, but failed to attract him at that salary. . . . I believe that this is an opportunity to secure a man with the best conception of the field of Geography, a man of ability to make the University of California known as one of the three main centers of Geographic research and geographic teaching among the Universities of the United States (*Holway to Barrows, March 2, 1920,* UA-PF).

Geography was a required subject in the College of Commerce which was headquartered in the Economics Department. The College of Commerce was not a true department but a course of study, its faculty composed of representatives from the departments teaching the required courses. Sauer was on the faculty of the College of Commerce when he came to Berkeley and, as Vice-Chairman, nominated himself Chairman in 1928-29. In 1943 the College of Commerce became the Department—later the School—of Business Administration. The association was important to the Department of Geography, as Commerce students filled its undergraduate courses. It was especially significant in the efforts of Holway to bring Sauer to Berkeley because there was a proposal to break up the department and farm out its courses to other departments. Stuart Daggett of Economics, then Chairman of the College of Commerce, wanted a strong Geography Department to take over the economic geography courses and supported Holway in his efforts to get Sauer.

Sauer was definitely interested in coming to California. He had written to Holway in April 1921, "If California offers me a satisfactory field to develop and a salary that enables me to carry my budget I shall be heartily glad to accept" (*Holway to Hatfield, April 7, 1921,* UA-PF). Legislative delay postponed action, and Sauer withdrew his name from consideration.

Holway, conscious that his retirement was approaching and that the Department was understaffed, continued to press for a new man. He wrote to President Barrows about departmental problems in October 1922. Notes on the back of this letter suggest the course of a consequent conversation. Under "Possibilities" is the remark that "Commercial Geography is well-taught at Mich. (Dr. Prof. Saur.)" [*sic*] and "Accommodations - Bacon Hall - 2 studies, and a laboratory - for maps models - Room 7 Basement. Geog. not necessarily accommodated with Geology" (*Holway to President, October 25, 1922,* UA-PF).

That December, Carl Plehn, Economics, was requested by H. R. Hatfield, Dean of the Faculties, to see a number of younger geographers, especially Carl Sauer, on his trip to the east. Throughout the spring, Holway continued to press for action, but the administration moved with its accustomed deliberate speed. On March 7, 1923, President Barrows telegraphed Sauer to offer him an appointment as Professor of Geography at a salary of $4,000. Sauer responded that the salary offered fell short of his "irreducible budget":

> The offer . . . raises most intriguing thoughts of what one might do in cultivating the virgin fields of the Far West and the Pacific area. . . . I have spent my time of apprenticeship at Michigan. I may transfer elsewhere to practice the trade I've been learning. I should be very glad indeed to approach the task of building at California a teaching staff, devoted also to productive scholarship, but my way is not clear unless I am able to maintain approximately my present income (*Sauer to Barrows, March 8, 1923,* UA-PF).

After some further prodding by Hatfield, President Barrows wired a cordial invitation to Sauer to suggest the salary and moving expenses for which he would be willing to come to Berkeley. "I am extremely interested in Geography myself. . . . I believe that you would find this a stimulating and congenial place to work" (*Barrows to Sauer, March 28, 1923,* UA-PF). He also approved bringing John B. Leighly (1895-1986) to Berkeley.

Sauer had previously written to Holway inquiring about the possibility of bringing Leighly with him, a man whose "intellectual curiosity is insatiable. I consider him the sort of man one finds only once in a blue moon." Sauer suggested that Leighly could work in the fields of cartography and physical geography, and help Sauer get his work established. Sauer wanted to specialize in regional geography, maintaining a permanent connection with the elementary work. The field of economic geography, in the sense of primary production and movement of trade, would require a third member of the staff (*Sauer to Holway, March 12, 1923,* UA-PF).

Sauer responded to President Barrow's telegram, asking for a salary of $5,000, senior rank in the department, and $500 for moving expenses, although it would cost him about $1,000 to move his household, library, and family. He also proposed teaching at least one summer session term for a couple of years so that he might familiarize himself with the local geography and the university. He wanted to develop a field course in central California similar to his summer work in Kentucky and Tennessee, and proposed to cover regional (human) geography permanently himself, continuing also the elementary work.

Sauer's proposals were accepted: a salary of $5,000, moving expenses of $500, two years of summer session teaching at $600 additional annually, and the expectation that he would develop the department as indicated. In addition, he was authorized to engage Leighly at $1,500 and told he might expect the active support and interest of allied departments.

Holway nominated Richard J. Russell (1895-1971), Senior Teaching Fellow, to be Associate for the coming year, as Burton Varney (1883-1943), who had taught climatology, was going east for an advanced degree. Russell would be the only member of the department with knowledge of its past work. He would complete his Ph.D. in geology in 1924. It turned out to be a felicitous association.

At Sauer's request, Holway nominated teaching fellows for the coming year, suggesting Alvena Suhl (1902-) for one position. Sauer soon "forgave her for being female" (PC-Storm) and reappointed her the next year. She received her M.A. from Berkeley in 1928 and, as Alvena Suhl Storm, developed the Department of Geography at San Diego State College and sent a number of excellent students to Berkeley for graduate work.

John Leighly describes the first hectic days on the Berkeley campus:

> In the interim between the retirement of R. S. Holway . . . and Sauer's arrival in August the geologists appropriated one of the three rooms previously used by geography. Desks and large cases for maps crowded the remaining two. Sauer immediately appealed to the president of the University for more space, which was promptly provided in the basement of South Hall, another of the original buildings, and which the department occupied for nearly ten years. Departments had the use of seminar rooms in the main library. Sauer found that F. J. Teggart, professor of social institutions, had sequestered the library's sets of bound volumes of the principal geographical journals in his seminar room. Getting them released was more difficult, and involved Sauer in more acrimony, than obtaining fit quarters for the department (Leighly 1976, p. 339).

Frederick J. Teggart (1870-1946) was something of a law unto himself on the Berkeley campus. He *was* the Department of Social Institutions. At the time of Sauer's arrival, he taught anthropogeography. Several years later he was teaching about the idea of progress in Western civilization. Sauer and Alfred Kroeber (1876-1960) were obliged to work with him in their various

73

schemes to develop a group major, to upgrade social science research at the University of California, and to achieve their own research goals. This was, however, an alliance of convenience, not affinity, as some of their later correspondence attests.

Sauer's first task was to revise the curriculum. Physiography (geomorphology), the basis of Holway's first course, had by agreement been given over to Geology. Sauer and Leighly together taught the introductory sequence. Geography 1, then as now, was "Physical Geography." Sauer and Leighly organized the course on the basis of climates and natural regions. Geography 2 embodied a new idea:

> A regional introduction to the geography of the world, in which the elements of geography are considered in their sum as differentiating the world into regions of different opportunities and limitations. These regions are then studied in terms of the character, density, and stage of development of population, with special recognition of regional unity of dominant economic systems or "cultures" (*Announcement of Courses of Instruction, 1923-24*, UA).

"Cultural Geography," as it came to be known, was Sauer's course for thirty years. (For a more complete description of courses and their development see Leighly 1976.) Both introductory courses were required by the College of Commerce. Sauer also taught Saturday and summer field courses, "Geography of North America," "Commercial Geography" (primary production), and "Principles of Geography" (geographic thought), using original sources in the original languages (Stanislawski 1975). In his second year he taught a seminar in "History of Geography." Sauer considered cartography an essential part of a geographer's training and, regretting his lack of skill, turned that instruction over to Leighly.

Sauer continued Holway's efforts to improve departmental facilities. He requested a classroom equipped to show projections, and a laboratory for the study of maps in the required sections of the introductory courses. Building up the map and slide collection was a significant budget item for many years. The department had received new instruments from the United States government for the weather station and these had to be properly installed. Because this station had one of the longest continuous records in the west, Sauer was able to keep the weather observer post in the department's budget through the Depression (UA-PF).

Geography was housed in South Hall with other social science departments—Economics, where the College of Commerce was headquartered, and Political Science. Sauer, in the course of faculty business and in actual propinquity, associated with social scientists. A number of men in these departments had a common interest in furthering social science research and worked with Sauer to establish the Institute of Social Sciences on campus.

74

The association of Sauer with Alfred Kroeber was long and productive. In the 1920s the Department of Anthropology was housed in the Museum of Anthropology in San Francisco, and Kroeber taught on campus one day a week. Leighly and Fred Kniffen (1900-   ) took Kroeber's seminar and may have initiated the acquaintance (PC-Leighly). The first communication of record is a formal request, November 9, 1925, from Kroeber to Sauer to review a manuscript. Sauer had found the concept of culture a valuable organizing principle before he met Kroeber, as shown by his use of the word in the course description for Geography 2, written before he arrived in Berkeley. It is likely that the two men enriched and deepened each other's ideas as they talked and worked together. They shared many interests and points of view, developed coordinated research projects, worked together in the field, and after several years taught joint seminars.

Both Sauer and Kroeber found productive possibilities for study in border fields, those at the edge, not the center, of a discipline. They favored joint or group research, and soon began to draw up coordinated research projects with Herbert Bolton (1870-1953) of History and Paul Taylor (1895-1984) of Economics. Coordinated research involving several fields was precisely the kind favored by the S.S.R.C. and the Laura S. Rockefeller Memorial Fund, as Kroeber no doubt knew from his association with eastern social scientists. Mexico and the American Southwest formed the geographical focus of these research projects.

Sauer may have first gone into Mexico in December 1923, in the course of a trip to San Diego to inspect high schools and normal schools accredited to the university. He may have crossed over to Tijuana, as most people did, according to Alvena Suhl Storm (PC-Storm).

In 1925 Sauer brought Oskar Schmieder (1891-1980) to the department to replace Richard Russell. Schmieder gives an amusing picture of Berkeley through a cosmopolite's eyes. He found Berkeley in the 1920s, especially the campus, clean, white, safe, and middle-class, where people left their doors unlocked. To the German, the students' manners were "sloppy" but not impertinent. He was struck by the conformity and prudishness of the community. The Dean of Women established standards of dress for women students. Young colleagues were shocked when he wore a red bathing suit without a top to swim in a Sierra lake, and people on the ferry laughed when he wore a straw hat before "Straw Hat Day." He remembered the seminars where professors, teaching fellows, and graduate students gathered around to discuss ideas or their work. He also enjoyed meeting men of other departments at lunch at the Faculty Club (Schmieder 1972).

Schmieder had been working in Argentina on settlement and agriculture. He became interested in Mexico after the seminar report by Peveril Meigs (1903-1979) and C. Warren Thornthwaite (1899-1963) of their trip into Baja California. Sauer had developed a notion that some California traits originated in Baja, "Mother of California," and encouraged the two students to make a reconnaissance in the summer of 1925. Intrigued himself, he took three students the following summer, May and June 1926, for a lengthy field trip into Baja California (West 1979).

Sauer found in Mexico the virgin fields he had hoped for. ''From year to year, that is, in the light of the last field experiences, more specific problems appear,'' he wrote in one of the numerous undated research proposals found in his papers, this one probably in 1928, before his 1930 sabbatical (SP). The Board of Research approved using Rockefeller funds to finance his first field excursions with graduate students ($600, not to include subsistence).

Institute of Social Sciences

The Bancroft Library became a nucleus for a group of scholars working on Mexico and the American Southwest. The group included Sauer and Kroeber, Paul Taylor, and Herbert Bolton, considered the ''Old Man.'' Before long this group of scholars formed themselves into an informal committee to discuss and coordinate their research. They developed extensive and long-term projects for work in Mexico and the American Southwest requiring more funds and a longer commitment than were possible through the Board of Research.[2]

On March 27, 1927, Kroeber appeared before the board to request its consideration and approval of a scheme of coordinated research into the problems of cultural contact in the American Southwest and Mexico. He planned to present it at the annual meeting of the S.S.R.C., and board approval would strengthen his case. The board was reluctant to approve Kroeber's proposal as Paul Taylor's investigation of Mexican migration fell within its scope. President W. W. Campbell's opinion was that Taylor's work was important and worth continuing, but, ''I desire the University to have no direct relationship with this activity. The subject is one which might seriously embarrass the University were the University in any direct way responsible for it'' (UA-BR).

In June, A. O. Leuschner (Astronomy), Chairman of the Board of Research, informed Campbell that Kroeber had been invited to meet with the Committee on Culture Areas of the S.S.R.C., indicating the committee's interest in the proposal. The S.S.R.C. responded negatively, observing that there was ''no organization in the University which could assume full responsibility for the administration of large funds allotted to it, a situation which might result in unfortunate jealousies among different departments'' (UA-BR).

The university was not organized to handle the kind of projects favored by the S.S.R.C. and the Rockefeller Memorial Fund, long-term projects cutting across departmental lines. Proposals had to go annually through the Board of Research. The president warned that the university could not be committed to a specified sum for a proposed activity each year; it would have to "run the risk of sharing in the University bounty" (UA-BR).

Charles Lipman, Dean of the Graduate Division, recommended that a Committee on Research in the Social Sciences be established as an administrative committee. Its function would be "recommendatory" on matters of research involving more than one department. The board voted to form a committee as Lipman suggested, to be made up of three ex-officio members and the department chairmen: Sauer, Bolton, Kroeber, R. G. Gettell (Political Science), Jessica B. Peixotto (Economics), and Teggart.

The committee set to work. One unsigned and undated proposal for an Institute of Social Sciences called for a million-dollar building and library, administrative and other expenses, and a half-million dollars annually for research. The institute was to focus on countries bordering the Pacific. Its purpose was the accumulation of facts "which can be properly correlated and interpreted to serve in the establishment of general principles for the control of human behavior and improvement of human relations" (UA-MA). Some of these ideas remained in the less ambitious proposal finally forwarded to the Rockefeller Memorial Fund. This included an outline of research possibilities along departmental lines. The proposed annual appropriation for all activities of the institute was to be $604,000.

Dean Lipman informed Edmund Day at the Memorial Fund what the university intended to contribute to the institute: in sum, nothing much. The university proposed to contribute the research activities of persons then active, the resources of the Bancroft Library, the collections of the Anthropology Department, the laboratories of Psychology and Geography, and grants for individual projects; and it promised to be liberal with leaves of absence (UA-PF). On October 2, 1928, Day informed President Campbell that the Memorial Fund could not "wisely undertake" to give the proposal more consideration. First, the sums ran much larger than the Memorial Fund could make available, and second, dependence on the Memorial Fund for so large a share of the financial requirements indicated that possible local support either had not been canvassed or did not exist.

Some sort of organization to coordinate research among the several social science departments was still necessary and advisable, however. On October 26 the committee submitted a draft plan of organization for an Institute of Social Sciences. Its purpose would be to further and to integrate studies and teaching in the field of social sciences. The plan called for an Advisory Council of fifteen members, including a five-member Executive Committee, both appointed by the president. On the copy of this plan that is in the president's file, someone (the president?) has crossed out the section that calls for the Advisory Council to examine and pass on projects for research and the suggestion that a modest emergency appropriation be made to initiate some research or other undertaking identified with the institute.

The Regents of the university authorized creation of the Institute of Social Sciences on January 9, 1929. Sauer was appointed to the Advisory Council; Kroeber, Bolton, Teggart, Ira Cross, and Dean Lipman were appointed to the Executive Committee. In his letter announcing the establishment of the institute, President Campbell stated that its purpose was the advancement of scholarship and research:

> It is intended that membership shall be open to those members of the departments of the social sciences who are seriously alive to the problems of the social sciences, who are contributing productively to their development, and who are deeply interested in improving the status of the social sciences in their various relationships. . . . The original idea arose from a common interest in a cooperative research project proposed jointly by Professor Bolton, Professor Kroeber, Professor Sauer, and Professor Teggart (UA-PF).

At the first meeting of the council, Lipman stressed the importance of presenting research proposals to the institute, and a Committee on Projects was appointed to make a survey. Carl Plehn (Economics), Teggart, and Sauer were appointed to this committee.

As the institute had its inception in the efforts of Sauer, Kroeber, and Bolton to get support for their joint research project in culture frontiers and culture hearths in Mexico and the American Southwest, Sauer continued to be active in its affairs. Establishing the institute was the response to the S.S.R.C.'s objection that the university was not organized to administer grants for large interdepartmental projects, and to that of the Rockefeller Memorial Fund that there was no local support for these projects. The task of the Committee on Projects was to fashion a plan or proposal as a basis for raising local and Rockefeller funds.

Sauer presented the committee's report on April 30, 1929. It stated that the purpose of the institute was to advance research, its first duty. It was to facilitate, to raise standards, and foster collaboration. It was to select and budget projects, and to solicit funds for and through the Board of Research (UA-MA).

This report reads very much like the work of Sauer, in both its approach and its wording. The activities are not defined in departmental terms, but geographically. The first area, California, an old focus of studies at the university, presented no problems with finding either subjects or money to support research. Some California scholars had been doing research on Pacific Hemisphere subjects. The University of California, Stanford University, and the University of Texas were the only institutions concerned with Mexico and the American Southwest, and here was a major opportunity for research inadequately financed. Another area of need was financial support for older scholars to enable them to do research away from Berkeley while on sabbatical.

Social science research at the University of California in those days, especially study away from campus, faced unique difficulties. The university administration showed little interest or understanding, and provided little or no support. There were problems related to a professor's temporary absence, such as funding of a replacement, and his or her status on return. Older scholars doing continuing research did not qualify for the fellowships and grants-in-aid available to younger students. With families to support, they could not afford to finance their own studies away from campus while on sabbatical. Conducting tropical or Latin American field studies presented special problems, as the best time for such expeditions is North America's winter. The summer in Mexico is too hot, useful only for archive work. The four- or five-week Christmas break was not long enough. University regulations did not allow granting funds for subsistence in the field.

The Committee on Projects proposed the following projects as most worthy of funding through the Institute of Social Sciences:

• Four stipends for research in Europe or the eastern United States.

• The coordinated research project in "Culture Frontiers and Culture Hearths" of Mexico and the American Southwest. There already was a nucleus of three departments, and several projects were ready to be undertaken.

• A Bureau of Economic Statistics, with a staff to collect the materials and help scholars use them.

The council approved the six-year program submitted by the Committee on Projects. The cost was $78,000 per annum, including research under way, far more than the Board of Research had at its disposal. Approval was given to start canvassing local sources of support before approaching the Rockefeller Foundation again. Campbell approved the effort to raise the money, but warned that the university could not be committed to making up the amount subsequently, when the six years were over. It was expected that Rockefeller would give $50,000 if the university could raise $25,000 annually. A year later, only about one-third of the needed amount had been committed, a proportion that held true for several years.

In 1932 the proposal was revised again after Kroeber had talked with foundation personnel in New York. A new joint project of research in Northeast Asia was added, and a proposed atlas of California. The Institute of Social Sciences was losing favor with the Rockefeller Foundation. If the University of California did no worse than other institutions, it did no better. The process of granting large sums for the institutions to administer as they thought best was not producing satisfactory results (*Kroeber to Sauer, 1932,* SP). The process of administering the fluid funds at Berkeley was cumbersome, requests going first to the Board of Research, then to the Institute of Social Sciences, next to the president, and back again. The "old guard" wanted to continue doing business as usual and not delegate any authority to a committee to select projects. The foundations were interested in new approaches, large, long-term, coordinated projects, such as the Mexican project proposed by the informal committee—Sauer, Kroeber, Bolton, and Taylor.

79

On the campus itself, the institute had some success in the mid-1930s. By 1931 there was a program committee and general meetings were held. By the spring of 1933 there were eighty members. Visiting lecturers were sponsored. There were periodic "smokers" to discuss problems. Sauer led one such discussion on teaching problems in social sciences in 1933. On the whole, the institute probably raised the level of social science research among the faculty and eased some of the difficulties. The institute existed at least until 1944 to screen and refer proposals to the Board of Research. By the mid-1930s Sauer had begun to lose interest in it, especially as it appeared that the administration and the Board of Research would not change their way of conducting business. More and more he devoted his energies to coordinated research projects and to working with scholars with similar interests, interinstitutionally as well as interdepartmentally.

Interest in Latin American Studies

Gottfried Pfeifer (1901-1985) had been based in the department as a Rockefeller Fellow while studying settlement and cultural succession in the Sonoma Valley. Sauer petitioned to add Pfeifer to the staff in 1930-31 to take over economic geography:

> If Geography is to participate in the associated and integrated studies of human activities that may some day deserve to be called Social Sciences, a student of economic geography is most needed . . .
> We are doing undergraduate teaching, graduate instruction, and research here with a smaller staff than is attempting all three anywhere else in the country. We have certain advantages of location and experience toward the development of a distinctive geographic school (*Sauer to Campbell, December 5, 1929,* UA-PF).

Sauer was to have sabbatical leave in the spring of 1930. There was no question that he would go to Mexico, but securing funds for expenses was a major problem. He had been in Sonora in summer 1928 and had found a wealth of inconclusive evidence. There were problems of the geomorphology of the Sierra Madre, of possible climatic change, of the extent of prehistoric cultures, and of the identity of existing Indians. "Between the American border and the Valley of Mexico there is a vacuum merely of knowledge, not of significant fact. We have peered into it enough at one corner to be convinced that here is another Southwest" where the great geographic tradition was established, and even greater ethnographic finds were made (*undated research proposal,* SP).

An ideal opportunity seemed to present itself when the Santa Fe Railroad acquired the Orient Lines and took an option on the Mexico Northwestern, in the very area Sauer was anxious to penetrate. Expenses exclusive of salary would be well over $12,000. He thought the railroad would have a natural interest in the resources along its new routes and drew up a proposal. The president of the university was to approach the president of the Santa Fe to ask for his support. Sauer expressed himself as anxious to try out in an area of fundamentally different economic structure something like the land utilization studies he had planned in Michigan, a regional inventory followed by a regional development plan (*undated proposal for study,* SP).

The Board of Research approved Sauer's projected field study and searched for means of funding in advance of the grant from the Santa Fe, which in fact never came. On September 20, 1929, after a summer in southeastern Arizona, Sauer withdrew the Santa Fe project temporarily and asked to substitute a field exploration of the West Mexican Corridor. The survey was part of the larger Southwest Culture Area program approved by the Institute of Social Sciences. The board voted $2,540 to Sauer, to be an advance to the institute. However, President Campbell did not think Sauer's grant should depend on uncertain institute funding, and granted the money from his emergency fund. The search for sources to commit funds annually for the institute's six-year program was not going well.

Sauer went to Sinaloa and Nayarit during his sabbatical leave. He coordinated his observations with Kroeber by letter and in the field, and with Pfeifer, who was also working in Mexico. Schmieder resigned while Sauer was still in the field:

> Schmieder's going puts us up before the problem of the Latin American field. There isn't an American geographer who knows anything about it. . . . There isn't a soul in this country who can work at all along Schmieder's lines. . . . It looks therefore as though it were up to me to cultivate this field. I'm pretty old to start being a Latin American geographer, but it may be up to me (*Sauer to Kroeber, March 1, 1930,* UA-MA).

In the same letter, he regrets that his time at some stops is too short to allow even a reconnaissance, but that he would not mind so much if he could be assured that the work would get done. ''I know now that I shall not be able to afford the luxury of a sabbatical again.''

The Committee on Culture Frontiers and Culture Studies—Bolton, Kroeber and Sauer, with the addition of Paul Taylor and with its scope now widened to include all of Latin America—continued to work on its coordinated research project, to attract other scholars interested in the problem, and to gain the approval of the Rockefeller people. Most of the support came from the Rockefeller Foundation. Indeed, however dissatisfied the foundation and the S.S.R.C. became with the University of California, they continued to find merit in the projects of Sauer and the committee. Sauer described the workings of the group to Stacy May, head of the Social Sciences Division of the Rockefeller Foundation:

> This working association of Latin Americanists has kept before it the general objective of culture history. We are not trying to bracket under one heading

conventional departmentalized studies that happen to be situated within the confines of Latin America. . . . We are trying to understand the fortunes of man in Latin America, considering that there has been a significant contiguity and continuity of man in that geographic frame. Perhaps I can take two common catch words and give them new juxtaposition, saying that we are interested in the personality of the culture or cultures of Latin America. This means that we have been working on studies concerned with the content of culture, with the values by which these cultures have lived, with the diffusion of cultural elements or the resistance thereto, with plasticity, stagnation, and self-destruction, in short with many questions of cultural growth and change (*Sauer to May, July 30, 1937,* SP).

In the above letter to May, Sauer relates how interest in Latin American studies developed on the Berkeley campus, starting with the formation ten years previously of the self-constituted Committee on Latin American Studies. The next step was to establish a new publications series, *Ibero-Americana.* They then set up a graduate program leading to the Ph.D. in Latin American studies; and finally the College of Letters and Science established an undergraduate major, Spanish-American History and Culture of California and the Southwest. A dozen or more scholars from many departments plus several from the University of California at Los Angeles had done work related to that of the original committee program. The letter to May summarized a conversation exploring the possibility of establishing a Latin American Institute, national in scope, which the group hoped could have its headquarters on the Berkeley campus. This national project was never realized at Berkeley. The Center for Latin American Studies is confined to Berkeley scholars.

Sauer described to May his own development as a Latin American scholar:

> I started in to Mexico to make an appraisal of its economic geography. I still hope some day to do this, but meanwhile I have opened up one trail that leads to another. . . .
> I soon saw that until I knew the Indian, his attainments and his values, I was incapable of making a critical study in that field. As a geographer I could no more disregard Indian culture than I can disregard climate in interpreting the forms of life. We have all become historians of culture, because the past is to an extraordinary extent the living present in Latin America.

On the recommendation of Kroeber, Sauer received a Guggenheim Fellowship to continue his field studies in western Mexico in the fall semester, 1931. His family did not accompany him into the field in Sonora. His ''Progress Report on Research Grant 243'' describes his experience:

> For the first six weeks in the field the temperature went daily to 100° and even to 108° and the nights afforded little relief. I was incapacitated for a time by being kicked in the chest by a mule. This happened just as we were about to start into the Sierra Madre and this trip had to be given up. For weeks I was unable to exert myself much. As I recovered my assistant's

digestion gave out, with the meagre and unvaried food which we had. He lost more than twenty pounds from a normally lean body, and finally developed a good case of jaundice. At this time also a vicious typhoid-malaria epidemic overran the country in which we were, aggravated by the fact that the population was half starved in that section. The physical circumstances of this field season were the worst I have ever encountered and our work suffered somewhat in consequence (*Progress Report, Research Grant 243,* SP).

He wrote to Kroeber from the field:

> I'm not very keen for the isolation and the food and the *mulas* of the back country. I'd as soon do committee work on campus, but I'll have to serve my stretch here. Has anything happened to the "new" series of publications or is it "*Alles bleibt beim alten*"?
> This is one hell of a country—politics like a hound's hind leg (*Sauer to Kroeber, September 13, 1931,* UA-MA).

Kroeber replied, "Your frame of mind about Latins is surprising only in that it has been so long coming. I reached it the second time in Peru," and added that it was especially hard if your wife was not along. The latest budget plan for the new series ought to work out very well, he thought (*Kroeber to Sauer, September 22, 1931,* UA-MA).

The Latin American group started to confer with the University Editorial Board in early 1931 about establishing a series. *Ibero-Americana* was authorized on December 5, 1931, with Sauer, Kroeber, and Bolton as editors. The first issue, *Aztatlán* (Sauer and Brand 1932), appeared several months later (see Parsons and Vonnegut 1983). By the end of the year, three issues had been published, two were in press, and five more in preparation.

University of California *Publications in Geography* and *Ibero-Americana* were the vehicles for most of Sauer's and his students' publications. He wrote to Kroeber from Arizona in 1934:

> I like this volcanic country fine and can see several good-sized morphologic jobs ahead provided one dare to proceed from assumptions non-Davisian, which is doubtful. I suspect that I and my youngsters could get none of our physical geography published anywhere in this country. But don't breathe a word of it, or we may not be able to publish at Berkeley. [Rollin D.] Salisbury [1858-1922] . . . died too soon, else we should have been spared this Council-of-Trent period. . . . he was the only man in the country who enjoyed a fracas with old W. M. D. and twitted the latter about having discovered the final scheme of things (*Sauer to Kroeber, July 10, 1934* UA-MA).

His work was out of the mainstream, indeed out of favor with most established geographers, and there was a general lack of interest in Latin America. "I am unaccustomed to any comment from other American geographer colleagues with the exception of my ever faithful band of Michigan friends," Sauer wrote to Isaiah Bowman (*September 8, 1932,* SP).

National Prominence

When he wrote the above letter to Kroeber, Sauer was in the field as advisor to the Soil Erosion Service (later Soil Conservation Service), where he had the opportunity to put to the test some of the morphologic ideas of soil-slope analysis that the members of the Geography Department at Berkeley had been developing. The students there had been introduced to the work of Walther Penck (1888-1923) from the time of Sauer's arrival, and Albrecht Penck (1858-1945) had visited Berkeley for a semester in 1925. Sauer had been pressed into service by the Soil Erosion Service after he had done a superb job of writing the Land Utilization Report for the Science Advisory Board (Sauer 1934a, 1934b). He wrote to Kroeber from New York in 1934:

> . . . the next direction in geography is going to come from the west. Leighly and I and perhaps Kesseli have got some things started that are, with the help of Russell, going to build a couple of bonfires under the physiographic mule, and part of the wood we're using is handiest out there. . . . The situation has stayed put for a generation in this country and it's due to break. Leighly's attacking on two fronts, and I on another. That's one reason I want to get out this summer. The program I've submitted will give our theses in soil-slope relations and in climatology the wider trial we've been looking for *(Sauer to Kroeber, n.d.,* UA-MA).

John E. Kesseli (1895-1980) had joined the department in 1932 to replace Pfeifer who had to return to Germany.

While Sauer's reputation was growing nationally, he was having troubles at home. The deepening Depression brought budget problems to the university. The Department of Geography was especially vulnerable because, although actively involved in research, it had not expanded its undergraduate offerings or its faculty. In March 1932 Sauer reported to C. B. Lipman on their objectives in graduate study and research, and concluded:

> We have been attempting to cover the range of geography, I believe, with a smaller personnel, more meagre support, and a more inadequate plant than is true at any American institution of rank. If the University of California is to exploit the uniqueness of its position and has confidence in our field and the work we are doing in it, there should be relief in all of these matters as soon as possible. We have made no bid for expansion by the means of building undergraduate enrollment; we do hope for recognition of the fields of investigation which may be ours for the taking *(Sauer to Lipman, March 14, 1932,* SP).

Sauer's idea of university education was that "barring certain orientation and disciplinary courses, a curriculum should reflect the serious intellectual preoccupations of the faculty" *(Sauer to Deutsch, March 23, 1934,* UA-PF).

Sauer had to fight hard to maintain the strength of his department during the mid-1930s. A policy in 1933 that there should be no promotions in rank hit

Leighly hard and drew the ire of Sauer. He wrote to President Sproul in protest:

> I have never discovered by what means departmental problems may be discussed with the Administration. . . . What the attitude of the Administration toward the departmental program is, or its knowledge thereof, I do not know, and our files are silent on that subject. . . . That I have spent ten years of steady, productive work with a single increase in salary which will be more than wiped out in the eleventh, hardly requires comment. Either my estimate of what I have been doing is in error, or the University is inflicting an unmerited penalty by the stand-still proviso. However deep a cut may be necessary, I think it should not be based on the assumption that the members of the University are standing still (*Sauer to Sproul, March 4, 1933,* UA-PF).

He wrote again to Sproul a month later, complaining of the massive cuts in his budget and comparing the department's position with those of other universities and with other departments at Berkeley:

> Of the other ranking geographic groups in this country, none is in as cramped and shabby quarters, none has as small a staff, none has received as little recognition from its university.
>
> At the time when I was invited to come to California, the President of this University informed me of the intention to support strongly the development of the work in Geography. . . . It would, however, be ungenerous not to lighten this retrospect by acknowledgment of obligation to the University in two respects: the University Press and the Board of Research (*Sauer to Sproul, April 8, 1933,* UA-PF).

The Chairman of the Budget Committee at this time was William Popper (Oriental Languages), who showed little appreciation of the Geography Department or geography in general. Especially, he noted that no consistent plan for the development of the department and no single theory of the content of its courses had been presented. "It is by no means clear that, aside from descriptive and informational matter, there exists any definite body of theoretical principles in terms of which advanced instruction in the department is at present organized" (*Popper to Deutsch, May 1, 1933,* UA-PF). He also suggested that because Sauer had been willing to come for $4,500 in 1921, and yet was given $5,000, he should be satisfied. The battle continued through the president's office through the spring of 1933 before there was any response from the administration. At least in one respect the situation of the department was relieved, when it moved to Giannini Hall in the fall of 1933.

The informal group of Latinists was now the Committee for Latin American Affairs of the Institute of Social Sciences and had attracted a dozen or more fellow scholars from Berkeley and Los Angeles. They continued to coordinate their research projects and their efforts to find financial support. According to Paul Taylor, Sauer "kept the kettle boiling" with paper organization (PC-Taylor). They pushed for the reorganization of the Bancroft

Library, and to augment the South and Central American collections. The group began to think of expanding beyond the boundaries of the university, to form a working association of Latin American scholars and, incidentally, be freer of the university administration jungle.

Sauer had been broadening his associations with scholars in other west coast universities through several organizations. In 1929 the S.S.R.C. established a Pacific States Committee to hold annual conferences. Carl Alsberg (1877-1940) of Stanford was chairman. He was a man after Sauer's own heart, a biochemist who became a leading economist, highly respected in the S.S.R.C. Sauer was appointed to this committee in 1932. In the same year he reported attending an annual International Relations Institute at Riverside to which the university sent several delegates, which he thought one of the more successful experiments in associating the various social sciences (*Sauer to Dodge, December 19, 1932,* SP).

About this time Sauer began to work on a new program for coordinated research in the Regional Economy of the Far West. This project would be not only interdepartmental but interinstitutional. Members included Carl Alsberg, chairman; Paul Taylor; and men from Reed and Pomona colleges. "Conversations . . . with members of the Rockefeller Foundation . . . indicate that both the inter-institutional character of the Committee and such a program as it contemplates will be watched with sympathetic interest" (*Request for Grant— Regional Economy of the Far West,* SP). The purpose of the organization was to be able to draw on the best scholars, without regard to institution or department. Sauer and Alsberg were each to request an equal amount of Rockefeller money from their universities.

Sauer was also active in the Institute of Pacific Relations, an international organization which, it was hoped, might sponsor the Institute of Latin American Studies. Major changes on the national scene brought the work of the Latin American scholars to national attention. President Roosevelt appointed Cordell Hull Secretary of State in 1933. Hull attended the Pan-American Conference in 1933, and developed the Good Neighbor Policy which focused attention on America's neighbors to the south. The S.S.R.C. and the A.C.L.S. set up the Joint Committee for Latin American Studies in 1933 and Bolton was appointed chairman. Paul Taylor and Isaiah Bowman were also members. The plan was to hold annual meetings. Sauer was invited to join the Committee on Latin American Research in January 1934. He responded, "I accept with pleasure . . . in particular since I was I believe the first local agitator for a periodic congress of Latin Americanists" (*Sauer to Munro, January 17, 1934,* SP).

In January 1934 Isaiah Bowman requested Sauer to write the report on land utilization for the Science Advisory Board. This was virtually a command and he had three days to change his semester plans (*Sauer to Wellington D. Jones, January 31, 1934,* SP). He was to arrive in New York early in February, where he would be based at the American Geographical Society.

Conscientious as ever, he had to complete his work for the committee for reorganization of the master's degree, give some attention to the Regional Committee, and go over three theses before leaving the next week. The difficulties would be with the administration. As he expected, the trivial things took far more time than the important ones (*Sauer to Bowman, Janury 26, 1934, SP*). Sauer proposed that the university continue to pay his salary and the Science Advisory Board pay that of his substitute. If the university suffers from the temporary absence of an experienced teacher, "it receives a compensating gain by having one of its members asked to apply his knowledge and time to a national question"(*Sauer to Deutsch, January 26, 1934, UA-PF*). Usually scholars from a nearby Atlantic Coast institution, rather than from the Pacific, were invited to perform this work. However, the administration could not see that Sauer was serving the university when not teaching. He was given leave of absence without salary, for three months.

Kroeber kept Sauer informed of events at the university. Everything in his master's program revision was voted down. The real reason was departmental autonomy, "the right of each to be as rotten as it pleased." His Regional Committee proposal had been referred to the University of California Regents. Sauer, in turn, kept Kroeber informed of his conversations with the Rockefeller people and of their growing impatience with the university and its cumbrous way of doing things.

While in the east, Sauer was asked to serve on the Advisory Committee for Population Redistribution of the S.S.R.C., established to advise the federal administration on problems arising from the movement of population. The first job was to be the Tennessee Valley Authority. The chairman was Joseph Willits (1889-1979) of the Wharton School of Economics. Sauer attended several of the committee meetings at Lake George.

Sauer's report on land utilization and the population studies received favorable attention, but at a cost to Sauer. He had received a Rockefeller grant to go to Central America in the summer to study the plantation system, but he had to postpone his trip to act as consultant for the Soil Erosion Service.

These were the heady days of the New Deal, when the Roosevelt Administration was experimenting on many fronts to solve the economic, environmental, and social problems that were facing it. New government agencies were set up, and planning and action were the order of the day. Sauer and Bowman shared some doubts as to the effectiveness of some of the programs, such as the Shelter Belt and the schemes to control soil erosion. Sauer felt, on the other hand, that the Population Committee had proceeded on the proper course of analyzing conditions as they are before programing future conditions:

I am . . . dismayed by the number of my friends . . . in the social sciences who are espousing what they call the field of economic planning and of social

87

planning. . . . I have had a couple of sad years of experience with planned control of soil erosion, on the basis of little or no knowledge of the causes of soil erosion. . . . I should hate very much to be considered as opposed to the use of science in prediction and program, but also I think that we know pitifully, tragically little about social and economic processes. I do resent substituting dynamics or theory for an organized knowledge of our social and economic conditions (*Sauer to Willits, December 9, 1935,* SP).

While Sauer was in New York, headquartered at the American Geographical Society, W. M. Davis died, and Sauer became contributing editor of the *Geographical Review.* "Thus the treaty of peace becomes one of amity and alliance" (*Sauer to Kroeber, March 17, 1934,* UA-MA). Sauer was not popular with the established geographical community, partly because his work was in direct contradiction to generally accepted theories and ideas, and partly because he was almost never able to attend the annual meetings, held at the time most suitable for his fieldwork.

It was not until 1939 that Isaiah Bowman, Glenn T. Trewartha (1896-1984), and John K. Wright (1891-1969) nominated Sauer for President of the A.A.G. Bowman expected that Sauer's reaction would be the same as his own had been, and added that Sauer had no more cause for complaint than he. "I know the forces back of the delay and the play of those forces only amuse me" (*Bowman to Sauer, February 14, 1939,* SP):

> If the boys are not so peeved at me that they will blackball me, I am quite willing to stand for the office of President of the Association of American Geographers. . . . I am grateful to you and the committee for considering that it is now my turn, and I think I can put myself into this thing in a constructive spirit (*Sauer to Bowman, March 2, 1939,* SP).

Finally Carl Ortwin Sauer was recognized by his colleagues as one of the most distinguished members of their profession.

Acknowledgments

I am deeply grateful to John Leighly, Elizabeth Sauer FitzSimmons, Alvena Storm, Paul Taylor, Robert West, and the staff of the Bancroft Library for their time and assistance.

Notes

1. The archival material is all in the Bancroft Library, University of California, Berkeley. The following abbreviations have been used for citations:
BP—Bolton Papers
SP—Sauer Papers
UA—University Archives
UA-BR—Board of Research Minutes
UA-MA—Museum of Anthropology (Kroeber Papers)
UA-PF—President's file
PC—Personal Communication
    John Leighly, conversations
    Alvena Suhl Storm, Letter, December 11, 1979
    Paul Taylor, conversation

2. The Board of Research was responsible for allocating research funds to all branches of the University of California.

Literature Cited

Fosdick, R. B. *The Story of the Rockefeller Foundation.* New York: Harper and Brothers, 1972.

Leighly, J. "Carl Ortwin Sauer, 1889-1975." *Annals,* Association of American Geographers, vol. 66, no. 3 (September 1976): 337-348.

Parsons, J. J., and N. Vonnegut. *60 Years of Berkeley Geography, 1923-1983.* Berkeley: University of California, Department of Geography, 1983.

Sauer, C. O. (with C. K. Leith and others). "'Preliminary Recommendations of the Land-Use Committee." In *Report of the Science Advisory Board, 1933-1934,* pp. 137-161. Washington, D.C., 1934a.

Sauer, C. O. "Preliminary Report of the Land-Use Committee on Land Resource and Land Use in Relation to Public Policy." In *Report of the Science Advisory Board, 1933-1934,* pp. 165-260. Washington, D.C., 1934b.

Sauer, C. O., and D. Brand. "Aztatlán: Prehistoric Mexican Frontier on the Pacific Coast." *Ibero-Americana,* no. 1. Berkeley: University of California Press, 1932.

Schmieder, O. "Lebenserinnerungen und Tagebuchblätten eines Geographer." *Schriften des Geographischen Institut der Universität Kiel,* Band 40. Kiel: Verlag Ferdinand Hirt, 1972.

Stanislawski, D. "Carl Ortwin Sauer, 1889-1975." *Journal of Geography,* vol. 74, no. 9 (December 1975): 548-554.

West, R. C. *Carl Sauer's Fieldwork in Latin America.* Ann Arbor: University Microfilms International, 1979.

Sauer South by Southwest: Antimodernism and the Austral Impulse

Kent Mathewson

The American South evoked a certain fascination in the eyes of Carl Ortwin Sauer throughout his long career as a place where the cultural past was everpresent and easily observed. Sauer also held a lifelong conviction that peoples, places, and livelihood patterns that offered opportunities for studying the conditions antecedent to modernity, especially those situations exhibiting a resistance to the forces of modernization, should be the focus of a humane geography. Despite this, the South never came to occupy a central place in his mature work and writings. Examination of his published work and unpublished correspondence suggests that, while he considered the American South a topic of considerable interest, it was one that he was content to let others investigate. Both his initial attraction to investigating southern themes and his later avoidance of this region as an arena of sustained research can partially be explained in terms of simple contingency. One also has to look at Sauer within the context of his times. These times stretched somewhat uncomfortably across America's transitions from a rural to an urban nation and beyond (Parsons 1976).

His regional interests were initially centered on his native grounds—Missouri, the Upper South and the greater Midwest. Among his earlier works, three publications dealt primarily with southern themes: his doctoral dissertation on the Ozark region (1920a), an article on the economic problems of the Ozarks (1920b), and his Pennyroyal study (1927a). As these are studies of areas best considered border areas between the South and Middle West, only in a peripheral sense was Sauer ever a student of the American South. However, the work of several of his students who wrote dissertations on southern themes, as well as associates and former students who investigated topics within a southern regional context, extended some of his earlier interests into selected areas of the South. Their research record also helps us to understand Sauer's relationship to the American South.[1]

90

Before he moved to Berkeley in 1923, Sauer's base as a graduate student of geography at the University of Chicago and his faculty position at the University of Michigan offered easy access to the areas of the upper South— first the Ozarks, then Kentucky—where much of his initial fieldwork was accomplished. Mere proximity to regions where he could exercise his early preference for studying landscapes on the edges of modernization does not fully account for his initial engagement with the South, nor his subsequent disengagement from it. During the period of Sauer's residence in the Midwest, he might have looked in the opposite direction, to the American North and the lands of the Upper Great Lakes where modernity's marks had only been selectively etched in the landscapes of resource extraction. Between these furrows, the ethnically distinctive folkways of recently arrived, far-northern Europeans presented challenges and opportunities for a geographer of Sauer's bent.

As part of his economic land survey work for the state of Michigan, Sauer ventured briefly in a northerly direction. The Finnish farmers and woodsmen whom Sauer saw engaged in subsistence activities in Michigan's Upper Peninsula interested him the most, precisely because their culture exhibited the strongest links to premodernity in the region (Leighly 1978, p. 119). More broadly, he might have ranged from the Great Lakes region into the boreal forests of Canada and beyond. One academic generation earlier, Franz Boas (1858-1942), the founder of North American anthropology, started out as a student of physics turned physical geographer and then laid the basis for the foundations of American cultural anthropology through his fieldwork among the Eskimos of the Canadian North (Trindell 1969, Speth 1978). Sauer's own successional history from geology student to physical geographer to chorologist to cultural geographer invites comparison. It was largely through interaction with Boas's students Alfred Kroeber (1876-1960) and Robert Lowie (1883-1957) at Berkeley that Sauer was to incorporate elements of Boas's conception of culture into his own vision of cultural geography.

The principal lessons that Sauer learned from his state-sponsored work in Michigan were both empirical and theoretical. The historical evidence he encountered in the formerly forested Michigan pine regions convinced him that the theme of "destructive exploitation" should become an important part of geography's agenda in future research. Perhaps even more importantly, the evidence of the legacy of "destructive exploitation" he witnessed in northern Michigan sealed his decade-long dissatisfaction with the environmental determinist perspective still widely accepted in geography at that time. Sauer realized that human agency was clearly a force that could overcome the constraints of "environmental influences," but in many cases human agency was also capable of obliterating both the natural environment and the limiting factors or influences associated with it, especially when channeled by complex politics (Kersten 1982). As Martin (1985) has pointed out, the fourth of five major "paradigm shifts" that occurred in North American geography in the past century was crystallized in large part through Carl Sauer's personal disenchantment with aspects of the Midwest, its geographers, and the theories that guided them.

After his move to Berkeley, his interests shifted in stages to Upper and Lower California, the American Southwest, northwestern Mexico, central Mexico, South America, Central America, the Caribbean, and finally the pan-tropical world at large. Though never entirely abandoned, his interest in the American South can be seen as a deferred agenda, one that might have engaged his attention but, for reasons of both contingency and deeper intent, did not. His career trajectory can be seen as a protracted and purposeful march both southward and back in time.

Sauer an Antimodernist?

In describing the antimodernist impulse, T. J. J. Lears (1981, p. 57) suggests that, "transatlantic in scope and sources, it drew on venerable traditions as well as contemporary cultural currents." These included: republican moralism, the revolt against positivism, and romantic literary convention. Republican moralists championed the virtues of the small farmer and local autonomy, while viewing the city as a scene where tendencies toward luxurious living could easily lapse into senescent overcivilization. It was also feared that once the urban scene became "overcivilized" it would disrupt the steady-state harmony of the middle landscape of virtuous small producers. Society as a whole would then be threatened with a descent into the irreversible organicist-cyclical historical patterns reminiscent of European and, even more ominously, of Asiatic societies. The revolt against positivism rejected both static and simplistic-progressive intellectual and moral systems, often in the name of a vitalist cult of energy and process. Romantic literary convention both diverged from and complemented republican moralism. The romantic vision elevated simple and childlike rusticity over the artifice and superficiality of civilization. While these three threads of the antimodernist sensibility are not wholly congruent, they stem from a single and simple source: discomfort with modernity.

The cement that united Sauer's regional foci with his varied enthusiasms was his own particular mode of antimodernism.[2] Sauer's conception of geography and its practice has been judged "idiosyncratic" by more than one observer including himself (Williams 1983, p. 2). It has not been easy to place Sauer's vision in a larger tradition or context, and hence much of his work has been considered divergent from both mainstream geography and twentieth-century social science. He was unself-consciously iconoclastic on many issues (Johnston 1979, p. 38; Entrikin 1984, p. 407) and appreciated the "maverick" tendency in associates (Leighly 1976, p. 344; 1978, p. 131). Hooson (1981, p. 166) aptly characterized him as "both a rock-ribbed conservative and a congenital nonconformist." The elusive nature of Sauer's work and vision might be better understood if placed in the company of other antimodernist culture historians and critics such as Henry Adams (1838-1918), Lewis Mumford (1895-   ), and T. S. Eliot (1888-1965), or the vitalist poet Robinson Jeffers (1887-1962) whose work Sauer admired. Uncomfortable with the modern world for differing reasons, antimodernists of widely divergent "political" positions and intensities emerged between 1880 and 1920 to leave a complex legacy of critiques of modernity.

Sauer's interest in the above and other prominent American antimodernists is documented in his published writings and correspondence. Without a study specifically directed at elucidating the nature of the influences that these and like-minded figures had on Sauer's intellectual formation and evolution, it is premature to say how strong the impact was. It is possible to point out a few tangible points of convergence. Sauer's affinity with Lewis Mumford and his ideas was direct and mutual. Sauer made this obvious by inviting Mumford to cochair the landmark symposium on "Man's Role in Changing the Face of the Earth" (Thomas 1956). While Sauer can hardly be accused of joining T. S. Eliot as a convert to religious and cultural Anglo-idolatry, the two self-exiled Missourians shared similar moral concern over civilization's abuse of the natural world (Speth 1977, p. 156). As regards Anglophilia, Sauer was more than just agnostic. At times he was antagonistic. This has been pointed out by Dunbar regarding Sauer's policy on visiting faculty:

> Sauer's predilection for German visitors did not extend to their insular cousins, as he said to Bowman in 1931, "The chipper little [English] schoolmasters, who from time to time demonstrate their teaching cleverness in this country, are out" (1981, p. 8).

The links between Sauer and regionalist, antimodernist poets such as Robinson Jeffers in California or the southern agrarians at Vanderbilt also await elaboration. Sauer expressed admiration for Jeffers's poetry from time to time. Both had fathers who taught languages and literature at colleges with theological affiliations. Both attended secondary schools in Germany. Both had an abiding respect for the German romantic tradition, especially Goethe's place within it. Both were attracted to the Spenglerian prophecy of the decline of Western civilization, while as immigrants to California they glimpsed redemptive potential for aspects of the American West (Coffin 1971).

It is also instructive to compare both the outlooks and the careers of Sauer and Henry Adams. By the beginning of this century, Henry Adams and his brother Brooks (1848-1927) had established themselves as widely read advocates of a pessimistic reading of American culture history. A year after Fredrick Jackson Turner's 1893 address to the American Historical Association announced the closing of the American frontier, Henry Adams in his presidential address to the same group announced the closure of both the Bucklean environmentalist and the Darwinian-inspired "progressivist" visions of history. He hinted at a devolutionary vision that took final form in his posthumously published *The Degradation of the Democratic Dogma* (Adams 1919). Sauer (1925, p. 31) cited Adams's programmatic sequel to the 1894 address, "The Rule of Phase as Applied to History," as an example of one of several "views of the structure of history" similar to Spengler's "mathematico-philosophical thesis of the culture cycle." He also asserted that Spengler's thesis was the "complete antithesis of Buckle . . . [it] is of such importance that it should be known to every geographer . . ." (Sauer 1925, p. 31).

Perpetrating a mild hoax, Adams evaded the duty of presenting his address in person by feigning absence and posting the paper to his colleagues with the fictitious dateline "Guada'-c-jara" [Mexico] (Stevenson 1958, p. 356). He suggested (in the text) that as it was being read he was "somewhere beyond the Isthmus of Panama," when in fact he was at home in Washington, D.C. Sauer may have considered following Adams's example with a genuine austral evasion of his presidential address to the Association of American Geographers in 1940 (West 1979, p. 94). The meeting was held in Baton Rouge, but Sauer complained of having to delay the start of his Mexican field season to attend. This emphasizes his regional priorities at this time; Mesoamerica took precedence over the Deep South.

While Sauer's and Adams's views on history and the world in general might have been similar, their personal histories were very different. Sauer may have been born into the world of small town America, but his own family setting was distinctively cosmopolitan in comparison to the class locations and geographical contexts of his neighbors and their surroundings.[3] The family traveled to Europe and his father had been offered and declined a faculty position at a "prestigious Midwestern university" (Kenzer 1985a). Though Sauer liked to speak of himself variously as a "provincial," a "peasant," and as a "Missouri farmer," these designations were more gestures of solidarity with the "homefolks" of his natal region than statements of fact. His multilingual and bicultural background made him simultaneously the accepted insider and sympathetic outsider (Kenzer 1985b). This quality of cultural dexterity helped him to move effectively within many cultural contexts in later life. Despite this maneuverability, Sauer was far removed from the circumstances that formed Henry Adams.

Adams's ancestral background included the accumulated weight of several generations of public service, social probity, and cultural accomplishments at levels ranging from local and regional to national and international. Adams could never place this looming and illustrious heritage behind him, but rather he wore it ambivalently like the world itself on his back. The tropisms this caused led in turn to his somewhat eccentric self-conceptions and constructions. For example, after abandoning his Harvard professorship in history during the Centennial Year 1876, he devoted himself for the next half-century mainly to writing history and offering advice to those in public service, with a tone that became increasingly pessimistic. At century's close he began a new round of foreign travels, including the Orient, Oceania, Latin

America, and Europe. The purpose of these trips (especially his tropical adventures) was not for public service, as his early trips to Europe had been. They were "austral flights" in a neo-Romantic or antimodernist sense. In Europe he consciously avoided the present by traveling on an aesthete's tour of medieval high culture landscapes. Sauer did something similar in his post-World War II visits to Europe. He sought out archives, the repositories of the past, and felt most at home in the relatively unchanged folkish landscapes of his ancestral southwestern Germany (Leighly 1976, p. 343).

One of the more important links between Sauer and Adams, especially for future attempts at specifying Sauer's outlook, is the importance both attached to curiosity in historical understanding. For Adams, as an aspect of human nature, curiosity was *the* motive force in history (Contosta 1980, p. 130). Time and again in Sauer's writings the importance of curiosity came out, often as justification for what he did as a geographer. It was also the standard remedy that he prescribed for those suffering from the "pernicious anemia" brought on by methodological questioning and insecurity (Sauer 1941, p. 4). Curiosity as the creative element in the human experience is close to the core of Sauer's conception of culture history. For example, Sauer basically argues that curiosity and not economic necessity was responsible for the processes that led to the domestication of plants and animals by humans. The optimum context for curiosity to flourish is not under conditions of economic stress; rather it is in situations where leisured life ways can unfold so that humane cultural innovations occur (Sauer 1952, p. 21). This perspective invites comparison with the culture theory expounded by the Dutch culture historian Huizinga (1950). Huizinga felt that the idea of *Homo ludens* rather than *H. faber* or *H. economicus* better represents the authentic nature of the way humans create their culture. Championing the *ludique* or play principle over the utilitarian interpretation of culture genesis has been one of the central themes in the culture conservative vision for the past two centuries (Weintraub 1966). Sauer's relation to this broad current in Western thought deserves a separate study.

Sauer and Republican Moralism

Sauer can be understood in terms of the essential antimodernist traits discussed by Lears (1981), as well as other traits that are related to the antimodernist sensibility. First and perhaps foremost, he strongly held the republican moralist position in regard to the virtues of yeoman husbandry of family and farm in the Middle West. Republican moralism refers in part to the perceived antinomy between public virtue and fortune and especially corruption, Fortuna's cohort in the postfeudal Atlantic world. In North America the debate is usually thought to have reached its apogee in the late colonial and early republican periods, after which Lockean liberalism replaced classical political theory as the dominant mode of political thought. In a sense Jefferson's agrarian republicanism was an attempt at a guerrilla holding action positing that virtue could be maintained, if at all, only in a rural or "extrapolitan" context. As the historian Popock has argued:

> There is thus a dimension of historical pessimism in American thought at its most utopian, which stems from the confrontation of virtue and commerce and threatens to reduce all American history to a Machiavellian or Rousseauan moment (1975, p. 541).

For Popock "the Machiavellian moment":

> . . . is a name for the moment in conceptualized time in which the republic was seen as confronting its own temporal finitude, as attempting to remain morally and politically stable in a stream of irrational events conceived as essentially destructive of all systems of secular stability (1975, p. viii).

Popock (1975, p. 506) sees the continuity of the Atlantic republican tradition on American shores as "being used to express an early form of the quarrel with modernity." Despite the dominant view in American historiography which suggests that the Atlantic republican tradition was superseded if not extirpated as a significant force in American intellectual and political life after about 1830, I would argue that it persisted as an important element in much of the antimodernist sentiment at the turn of this century.

Sauer's own political, or perhaps more accurately, antipolitical vision was a reformed variant of republican moralism. This vision was strongly informed by his concern for ecological or environmental stability. As voiced by Sauer, the classical republican concept of virtue was to be realized in sane stewardship of the land. It could also be expressed through historical scholarship that exposes the corruption of virtue that arises from modernity's destructive exploitation of nature. Fortuna's confederacy with commerce in the modern age made the prospects for ecological stability remote at best.

Sauer and Positivism

Sauer's contempt for positivism was a lifelong commitment. For many, this position represents the most enduring testimony to his profound prescience and common sense. His work countered positivism in three of its main expressions. First, in Latin America he defended the aboriginal cultural heritage that local elites were attempting to eradicate through "progress" in the name of Comtean positivism. Second, he confronted the North American boosters of the doctrine of order and progress by uncompromisingly denouncing mindless "development" and the alleged benefits of modernization. In turn, he subtly demonstrated the erroneous and even banal assumptions embedded in much of the modernization theory that began to emerge in the 1930s and came to assume a certain hegemony in development studies in the post-World War II period. At the same time, Sauer's critique of developmental theory and policy offered an important alternative to most leftist positions which attack modernization theory while ignoring the ecological constraints that impinge on social theorizing. Finally, during the divisive decade when sectarian ''nomothetes'' waged a campaign of modernization within geography itself, Sauer and his associates offered a counterbalance to the more *il*logical positivists among the ''revolutionists'' (see Billinge, Gregory and Martin 1983).

Antimodernism and the State

Sauer's consistent stand against positivism invites speculation regarding his political outlook. One of the more perplexing aspects of Lears's (1981) discussion of American antimodernists is the array of political postures implied. At first glance there seems to be only diversity. The only unity visible is a common distrust of modernity. However, if these representative figures are considered in relation to their attitude toward the state and differing social formations, patterns emerge.

Most of Lears's *dramatis personae* viewed medieval feudalism, in both its occidental and oriental modes, as particularly appealing (cf. Mumford 1922, 1961; Wright 1925, 1966). This befits the elite social origins and high culture enthusiasms of many of these figures. Standard traits attributed to medieval feudalism include a weak state structure complemented by an idealized spirit of deference to ascriptive social hierarchy, and an ''organic'' culture including a tradition of artisanal excellence in the realm of production. These features offered alternatives to the centralizing forces associated with technology and mass society under an industrial capitalism recently triumphant. Sauer, however, displayed little interest in medieval Europe or the Orient. Perhaps his most sustained engagement with the medieval theme was in writing his book *Northern Mists* (1966). Yet this is a study of hardy individualists living mostly on or beyond the Celtic fringe, setting forth to breach the Hesperides and concomitantly medieval constraints. As such, it is simply one possible prologue for the story of their more ''sedentary'' descendants who came to rest in the American Midwest.

Sauer has more in common with those antimodernists who expressed respect for social formations wherein a weak or nonexistent state structure is underwritten by a rural-based egalitarianism. Thomas Jefferson's agrarian republicanism, on the one hand, and the idealized communitarianism of Neolithic village life, on the other, can be viewed as points on a single temporal continuum. In the minds of those enchanted by this vision, the documented record of the latter offers support for the implementation of the former. With the closure of the North America frontier, both physically and metaphorically, early in his boyhood, Sauer probably realized that Jefferson's

97

ideal had paradoxically become more inaccessible than its remote precursor. This is perhaps one explanation for Sauer's incessant search for the ever deeper antecedent conditions of cultural phenomena. In effect, it was an inversion of Midwestern progressivism. The romantic German culture historical ideal of the search for the *"Ur"* principle replaced the dogma of progress as the guiding light in scholarship.

One does not have to travel far back in human history before the state, either seen as a progressive development or as a repressive construct, simply evaporates. Some antimodernists find comfort in the fact that humans have been "stateless" beings for the vast majority of their collective history. There are parallels, and in places convergences, between various antimodernist visions and those of libertarians and anarchists. This is particularly true on the issue of the state. Anarchists often appeal to this *Ur*-history to argue the unnaturalness of the state. One of the most articulate defenders of this idea was Peter Kropotkin (1842-1921), anarchist theoretician, revolutionist, and geographer. Various figures sharing elements of the antimodernist vision have drawn on Kropotkin's left-libertarian rendering of history, anthropology, and biology to support their critiques of modernity, including Mumford (1934) and Patrick Geddes (1854-1932) (see Geddes 1915). Kropotkin can be also included in the ranks of the antimodernists, but with certain qualifications.

Had Sauer sought to put his antistatist and antimodernist perspectives into the political terms associated with anarchism, we might see an affinity with Kropotkin. While Sauer (1936, p. 278) cites Kropotkin's scientific work, there is little to suggest that Sauer was interested in the plans of Kropotkin and other anarchists for replacing the state with various forms of extreme democracy. It should be noted that, in mentioning the "brilliant writings" of Kropotkin's fellow anarchist geographer Eliseé Reclus (1830-1905), Sauer (1927b, p. 156) did aver that Reclus was "excluded from an academic career because of political non-conformity. . . ." Sauer seemed equally uninterested in placing himself in the tradition of libertarian individualists on the "right" side of the political spectrum.[4] However, he did express contempt for the state as an institution on many occasions. In a rare instance where Sauer seemed to grant the state a certain legitimacy—in mobilizing for World War II—he still gave it only begrudging due:

> The longer the emergency, the greater the tendency for the State to dictate the values of life, to wipe out the distinction between things that are Caesar's and those that are God's. In the end, Caesar becomes divine and the republic is lost. Perhaps our greatest danger is that we who wish to destroy the tyranny of the state Moloch become idolators ourselves. . . . For the present we may adopt the view of H. G. Wells that the State is, in principle, a combat organization. Its business is to prevail over others (Sauer 1932, p. 10).

Interestingly enough, this meditation was written on New Year's Day, 1942, in the "stateless" space of the high seas south of Cuba. Sauer never did venture a formal critique of the state, but it appears to have been one of the principal and persistent irritants among many that he associated with the modern age. His quarrel with the state was something that he did not bother to conceptualize in formal terms. This was in line with his position on theory in general; he claimed to have no use for it. At best, even generalizations were meant to be transitory and provisional, to be evoked in the manner he thought hypotheses should be used, as "working hypotheses" and little more (Entrikin 1984, p. 391).

Henry Adams, perhaps *the* emblematic end-of-the-century antimodernist, referred to himself as a "Conservative Christian Anarchist" toward the end of his life, though it is uncertain what importance he attached to the label (Contosta 1980, p. 137). One can scarcely imagine him meaning it in the sense that Leo Tolstoy devoted himself late in life to becoming a Christian anarchist and an "honorary peasant." Sauer's own pietistic heritage, pacifist leanings, and communitarian sympathies, combined with his dislike of bureaucracies and the state, might place him near the political terrain where anarchists have most often congregated (Hewes 1983, p. 145). His principled nonconformism probably sets him apart from even this heterogeneous collectivity. To borrow Williams's (1983, p. 2) evocative image, Sauer was like an "intellectual Voortrekker" breaking camp and moving on at first sight of another man's epistemological smoke.

As a final aspect of his stand *contra* positivism, Sauer enthusiastically promoted what Lears (1981, p. 57) terms the antimodernist's "vitalist cult of energy and process" in his advocacy of fieldwork, or direct experience and engagement with one's object(s) of study. In Sauer's case, the focus was primarily on rural landscapes and the people who modified them. The concern for process can also be noted in his advocacy of genetic or historical approaches to human geography. His concern with process can be linked to his training in and by geomorphologists (Entrikin 1984). Adapting this processual perspective to an emergent cultural geography was one of the more important contributions Sauer made to the subdiscipline he pioneered.

Romanticism and the Austral Impulse

Sauer was never an unrestrained adherent of romanticism, especially when at home base in his capacity as a scientist and scholar, either in the Midwest or later in California. Nevertheless, his capacity for romantic celebration of the simple and the rustic seemed to increase as he traveled farther afield. Like Johann Wolfgang von Goethe (1749-1832), whom Sauer greatly admired, his romantic tendencies were both bounded by and grounded in the study of natural history informed by a morphological perspective (Speth 1981). Sauer also shared with Goethe an affinity for the "sunny south lands." The force of attraction for southern climes and landscapes on the romantic imagination might be termed the "austral impulse." Goethe experienced an early midlife renewal through his flight from the constraints of court life in Weimar and his personal discovery of both the classical past and a southern folk life in the Italian landscape (Goethe 1962). Alexander von Humboldt (1769-1859), close friend and intellectual kinsman of Goethe, had a similar objective in mind when he abandoned his executive position in the Prussian mining bureaucracy to join the Napoleonic scientific expedition to Egypt. Frustrated by the contingencies of war in the eastern Mediterranean, Humboldt embarked on his own monumentally successful Enlightenment venture and Romantic voyage to America's Mediterranean and its equinoctial regions (Humboldt 1852).

Sauer, after marking time in Michigan involved in somewhat more mundane bureaucratic duties, firmly established a career trajectory that took him south via the American Southwest to landscapes where remnants of the New World's "classical antiquities" were still evident. In Mexico and farther south, Sauer backtracked along many of von Humboldt's own trails. However, unlike either von Humboldt or Goethe, Sauer was less interested in the forms of the past frozen in stone than in seeing survivals of a more distant past alive in local folkways. Nor did Sauer ever develop much interest in investigating the high culture aspects of ancient New World civilizations. To the degree that the project of the Enlightenment had become the impoverished property of positivism since Goethe and von Humbold's time, Sauer's compass was set south by a somewhat different romantic reading. But like Goethe and von Humboldt, Sauer was revitalized by breaking out of an institutional and intellectual context that had become a burden.

What I have called the "austral impulse" is but one aspect of the romantic current within Western intellectual and social history since its inception in the eighteenth century. It should be seen as one of the more luminous or perhaps refractory strands within the total tapestry of thought and action that forms the backdrop for Romanticism in its historical and experiential contexts.

The early Romantic travel urge was, perhaps, an individualistic successor to Europe's absolutist efforts at global conquest and exploration. Reflecting its own etymological antecedents, Romantic travel was often work, but the labor yielded rewards of a personal, experiential nature—not the kind that could be materially accumulated. Twentieth-century tourism, at the level of appearances at least, can be viewed as a dissipative inversion of the earlier European thrusts toward conquest and exploration. The modern tourist is in part the inadvertent offspring of the high-minded Romantic traveler. The austral impulse has been "democratized," institutionalized, and above all valorized. It is true that solar energy as a source of power continues to evade efforts to monopolize it, but it has long since become a popular commodity when packaged with the appropriate setting. As Paul Fussel has shown, the vogue for solar consumption and southward travel has its own history. Immediately after World War I, or at the juncture when tourism supplanted travel (see Fussel 1980, pp. 37-50) as a dominant mode of landscape consumption in the North Atlantic world, Victorian "heliophobia" gave way to a new wave of sun-seekers. Fussel documents in his study of British literary travelers between the wars that the Romantic attraction for Mediterranean landscapes experienced a revival during this period. The same thing happened in North America, as California, the American Southwest, and the Gulf and Caribbean coasts experienced new popularity among literary travelers as well as tourists. One only has to point to D. H. Lawrence's (1926, 1928) panegyrics to premodern landscapes in the American Southwest and Mexico or Aldous Huxley's (1934) antimodernist rambles in *Beyond the Mexique Bay* to see the same occurring on this side of the Atlantic.

While Sauer's attraction to California and Mexico had many elements that distinguish it from the prevailing vogue in literary and popular culture, it would be foolish to view his affinity for these regions outside of the context of the times. As West has written:

> Several factors may have prompted Sauer's initial interest in Mexico and things Latin American. During his first years on the Berkeley campus he and his family usually visited southern California during the Christmas holidays, chiefly to escape the fog and chilly winter rains of the Bay Area (1979, p. 29).

This may seem so obvious and commonsensical to the contemporary reader that it should not merit comment. Fussel argues that during the late Victorian period, in England at least, there was a stay-at-home mentality even among those who could afford to escape the winter weather for warmer and sunnier climes: "Before the war one had been rather proud of the fogs and damps and pleased to exhibit staunchness and good humour in adapting to them" (Fussel 1980, p. 133). The 1920s occasioned something of a "solar revolution," and Fussel calls it "one of the most startling reversals in modern intellectual and emotional history." The heliophobia of the Victorian colonialists, "an indispensible accessory of the class system, at home distinguishing the fine from the less fine, in the colonies demarcating administrator from underdogs," gave way to its obverse, "the new heliophily" (1980, p. 138). Its manifestations ranged from the sun worship of the German back-to-nature movement to the core iconography of the Art Deco craze, to the real estate booms in California, Florida, and France's Côte d'Azur (Gade 1982). Thus going south had a certain cultural currency to it in the 1920s that has persisted until the present, but with different and diminished significance.

Sauer's New Middle Border

As Lears argues, the primary locus of American antimodernist sentiment was in the centers of patrician culture in the American Northeast. Sauer, by virtue of his German ancestry and schooling and his upbringing in a German-American area of Missouri, did not share the same formative experiences as the Henry and Brooks Adamses or the Lewis Mumfords of American cultural criticism. In terms of background, Sauer was closer to the Iowan, German-American Aldo Leopold (1887-1948), with whom he shared a Thoreauvian outlook on the American scene. Sauer, however, remained more closely in touch with the Midwestern experience than Leopold who became acculturated to the norms of the Northeastern elite through his schooling at Lawrenceville, Yale, and Yale's School of Forestry (Flader 1974). Sauer's origins and outlook also could be compared profitably to the greatest Missouri-Iowan critic of American culture, Samuel Clemens (1835-1910). As a young man living in Iowa, inspired by the Herdon-Gibbons expedition a few years earlier, Clemens took his first trip down the Mississippi in 1856 with plans to establish a coca plantation in Amazonia. He turned back at New Orleans (Smith 1950, p. 177). Later, writing as Mark Twain, he planted the notion of an "escape to Indian territory" in the mind of Huck Finn as a solution to Huck and Jim's Midwestern and Southern dilemmas.

If Sauer expressed blunt opinions about the uninspired and make-work plodders of geography he left behind in the Midwest in the mid-1920s, he was also capable of caustic asides on the overcivilized Brahmin professoriate in their eastern establishments. He forged for himself and his students a "middle border" beyond the pedestrian Midwest and a continent's breadth from the ascriptive realms of prestige and entitlement of some eastern institutions.

The middle border that Sauer sought to cultivate opened onto the south from his Berkeley vantage point. The old east-west axis linking an urbanized east with a westwardly receding wilderness was transformed into its axial counterpart. The new axis assumed a north-south alignment, with the civilized pole in California and opportunities for wilderness experience and exploration to be found in the equatorial tropics. The northern Euro-American landscape continuum mediated by postpeasant, smallholders with distant but distinguishable Neolithic roots was replaced with an Hispanic-aboriginal construct. Along this southward trajectory the farmers were still peasants, the Neolithic was much nearer, and the remains of ancient civilizations were clearly evident. Thus the American Southwest, and by extension Old Mexico, but not the American South, became Sauer's new middle border (Parsons 1979, p. 14).

Sauer's Southern Work

In bypassing the American South, Sauer coincidentally avoided confronting the weight of the recent political economic past. He never directly engaged the issues of slavery and production modes oriented toward either smallholder-subsistence or plantation agriculture in the antebellum American South. Sauer's most sustained commentary on the South came during the 1930s and 1940s following his Soil Commission work and travel to the Piedmont areas to survey landscape change (Williams 1983, p. 8). By this time, the Ozarks and the Pennyroyal had become simply citations to his earlier work.

In all, Sauer spent perhaps some twenty months of field time in the American South if one includes his postretirement visiting positions at campuses in or near the South, such as Louisiana State University, the University of Texas, and Indiana University. These and other brief visits allowed side trips into the southern countryside. Of course, he had also spent his boyhood on the edge of the Ozarks. As he remarked in the preface to his dissertation on the Ozarks:

> . . . [it] is the outgrowth of long acquaintance with the area and of deep affection for it. It is, in fact, a study in home geography, a study of the old home with its many and vivid associations (1920a, p. viii).

Of directed or systematic fieldwork, Sauer spent part of the fall of 1914 and the summer of 1915—perhaps some five months—in the South. He also spent the summer of 1917 teaching at George Peabody College in Nashville. From 1920 to 1923 Sauer spent part of his summers in Kentucky as director of the University of Michigan's Mill Springs Field Camp (James 1983). In 1919, or perhaps earlier, he did reconnaissance in southern Kentucky to locate a site for the camp. He spent much of the summer of 1923, his last as a resident of the Midwest, in the field working on the Pennyroyal study (Sauer 1927a) with students Clarence Newman and John Leighly.

An interval of more than a decade elapsed before his next southern exposure. In this intervening period he had greatly expanded his horizons, both intellectually and geographically. He had discovered the American Southwest and Mexico, and was beginning to be drawn increasingly toward the earliest evidence of human time and into the humid tropics. His return to the "humid east" was as a soil conservation consultant. On a Christmas trip to Missouri in 1934, he continued on to the Piedmont of the Carolinas and Virgina en route to Washington, and was shocked by the extent of soil destruction wrought by two centuries of abuse. In the wake of this trip, Sauer wrote to the regional sociologist Howard W. Odum: "From the standpoint of social science studies, I have renewed my old feeling that the Old South is the most exciting part of the United States" (1935a). In a subsequent letter Sauer expressed alarm at what he had seen. It is tinged with a pessimism that Odum probably agreed with:

. . . I am still shocked by what I saw of soil erosion over that portion of the Piedmont which I roamed. . . . most of the country I saw literally faces extinction of its economic structure in about one generation. We can muddle along out in this part of the world [California] and still make quite a few mistakes, but there is a good deal of the South where the penalties likely to be incurred are pretty staggering (1935b).

Both Odum and Sauer shared a measured appreciation of the Spenglerian outlook (O'Brien 1979, p. 66). The South's lack of urbanization offered potential redemption, but the legacy of plantation monoculture, especially in the Piedmont, made the outcome uncertain at best.

Sauer wrote two brief but important papers in 1938 on the theme of plant and animal destruction within the purview of economic history for public presentation. In both addresses he cited the southern record of soil destruction under the aegis of plantation production as particularly instructive and disturbing (Sauer 1938a, 1938b). Sauer's fusion of antimodernist pessimism and populist polemic was particularly sharp in the paper written for the Twelfth International Geographical Congress in Amsterdam. He resoundingly condemned colonialism for its domination and destruction of lands and livelihood patterns on the peripheries of the modern world. His paper stands out amid the generally self-congratulatory tone of reports by the world's leading geographers on the progress of the civilizing mission of Europe within its colonies. His recent work in the South figured prominently in his examples of the consequences of monoculture and modernization in agriculture and resource extraction.

Sauer turned again to his interests in the New World tropics in the late 1930s, after this midcareer reacquaintance with the South. In the 1940s he traveled beyond Mexico to South America. In the 1950s he visited and directed students' work in the Caribbean. Observations on his sporadic travels in the American South, especially after his "retirement" in 1957, do not account for more than a few lines in any of his published writing. The exception was his article, "Status and Change in the Rural Midwest—A Retrospect" (Sauer 1963). Here he focused on the changes that had either transformed or bypassed his early haunts. Most of the changes, to his eyes, had been for the worse. As he remarked, "The cities have taken over the countrysides. What was functionally good then has been largely replaced by other ways of life . . ." (1963, p. 365).

He did find cause for optimism, however, in some of the interstices of the process of modernization. For example, Sauer (1963, p. 361) noted that the wilderness was moving back into areas where small farmers had been forced out only a decade or two before. Land abandoned in the wake of wanton mineral extraction, or released from an agriculture so bound to the isotropic plane that it could scarcely tolerate natural relief, represented places where wildlife and successional vegetation could once again prosper.

Among Sauer's six students who wrote dissertations on southern themes, and the additional half-dozen who wrote articles on southern topics, most emphasized the distribution and diffusion of material culture traits. Themes that Sauer cultivated in his Latin American works such as the processes of plant and animal domestication, aboriginal demographic collapse, and landscape modification under the impact of European colonization did not figure prominently, if at all, in the work of his students on the South. Nor did they take up the challenge of investigating the region's legacy of slave and plantation economies. The themes they focused on included vernacular architecture, settlement patterns, ethnic isolates, and relic folkways. While these themes have been established as major concerns for cultural geographers, evincing ruralist rather than urbanist biases, they are less amenable to processual approaches than were many of Sauer's pursuits involving questions of prehistory, biogeography, and landscape evolution.

Sauer's "Austral Impulse": An Antimodernist Apotheosis?

Sauer reserved his full powers of intellect and intuition or what might be termed "retrospective prophecy" for neotropical America. He directed most of his students to this area as well. Perhaps what these lands promised more than anything was a passage to premodernity. For the Atlanticist antimodernists such as Henry Adams, the medieval landscapes and cathedrals of Europe offered temporary refuge and opportunities for renewal. For classmates of Adams such as Theodore Roosevelt, pursuit of premodern experience led variously to the North American frontier and wilderness or to martial adventures in the Caribbean and explorations in Amazonia. These were forms of antimodernism that Sauer had few if any affinities with, especially Roosevelt's "progressive" Latin American military interventions.

Sauer's contact with premodern peoples and conditions began in a sustained way in the 1920s, after the initial antimodernist "moment" as described by Lears had subsided. Moreover, interest in specifying the attitudes and sentiments that underwrote Sauer's highly regarded scholarly efforts has only recently become a subject of study. His writings, personal correspondence, and recorded statements all point to a complex sensibility that can be characterized most effectively as antimodernist. He shared with other prominent culture historians and critics of the past century a profound skepticism concerning modernity. That his critique of modernity was based on more than the simple *ressentiment* of the beleagured defender of rustic values is quite clear from his writings. This critique of modernity rests squarely on the proposition that the putative capacity of modernist social science for ameliorating the human condition is at best sophistry unless it leaves open the possibility of a contemporary science based in the collective cultural wisdom accumulated through all the millenia of human/environmental co-evolution.

105

Sauer found scant evidence that his cultural-historical/ecological vision could be accommodated either by academia or modern society at large. Thus he remained antimodernist until the end, championing an ecologically sensitive variant of republican moralism, persistently placing himself and his work in opposition to the strictures of positivism in its various modes, and often writing evocative prose resonating in places with Goethe's own natural historical writings. He looked to the older South beyond the southern United States less for personal or cultural salvation than for folkways and cultural survivals inviting study and undertstanding outside the bounds of modernity's destructive marches.

Acknowledgments

I wish to thank William M. Denevan, Yi Fu Tuan, and David Ward for discussions concerning this topic. Respectively, they contributed: Berkeley lore, interest, and expertise; winsome and well-considered comments; instant and incisive intellection. Fellow students Richard Mahon and Pascal O. Girot read and listened to work in progress. Kin and kindreds in different places both south and west provided welcomed advantage points from which some of these ideas were set in motion. I would also like to thank the directors and personnel of the Southern Historical Collection at the University of North Carolina, Chapel Hill, for their aid and for permission to quote from portions of the Howard W. Odum-Carl O. Sauer correspondence in their collection.

Notes

1. Sauer's students who wrote doctoral dissertations on topics set in the South or its border areas include: Thornthwaite (1930), Dicken (1931), Post (1937), Hewes (1940), Price (1950), and Zelinsky (1953). Other Sauer students and Berkeley School associates have done important work on southern topics, especially Kniffen (1936) and Leighly (1963). A more complete accounting of this work is in preparation as "Sauer and the South: A 'Deferred Agenda.'"

2. The term "antimodernism" remains ill-defined. I am indebted to David Ward for first suggesting that Sauer's attitudes and outlook could be in part described as "antimodernist," somewhat before its recent use emerged to describe a variety of positions and intellectual postures in differing disciplinary contexts. T. J. J. Lears (1981) uses the term in several general senses to advance his reading of American cultural history. Unfortunately Lears does not offer a critical discussion of the concept *per se*, beyond the one that emerges from his study of currents within American culture that provoked quarrels with modernity among certain prominent figures at the turn of the century. Central to Lears's discussion of the antimodernist impulse is the quest for intense forms of physical and spiritual experience as an antidote for the banality of mass societal norms then emerging with early corporate capitalism. Lears interprets much of the literary work within the modernist genre as antimodernist in intent when viewed on a cultural historical level. For a more precise interpretation of modernism, antimodernism, and postmodernism in the literary context see Lodge (1977). For useful discussions of these same terms but from current social-theoretical perspectives, see Giddens (1981), Habermas (1981), and Lyotard (1984).

Finally, the term antimodernism has begun to appear recently in journalistic commentary concerning a wide range of cultural trends implicated in the discomfort with modernity, e.g., Kramer (1984). Kramer views the resurgence of the *Heimat* sentiment (roughly a conflation of "home" + "homeland") in Germany as symptomatic of the antimodernist outlook shared by both political conservatives and segments of the Green Party in Germany today.

3. In his correspondence to W. P. Webb, Sauer made a fittingly ironic allusion to both his intellectual proximity to and social distance from Henry Adams. He remarked that as "a kid from the country, studying at the University of Chicago" he was at first taken in by the urbanity and apparent wisdom of his academic elders. They were entranced at the time by Bull Moose progressivism and its "millenial" [*sic*] calling. But after 1914 when "the world blew up. . . . The slow and painful Education of Carl Sauer had begun" (Williams 1982, p. 21). One might also find allusion to Adams's autobiographical-philosophical tract, *The Education of Henry Adams* (1918) in Sauer's honorary presidential address to the A.A.G.(1956).

4. Trying to place Sauer's social-philosophical stance in political perspective by drawing on the narrow range of categories and conceptions available in normative American political theory and discourse predictably identifies him as a "political conservative." Sauer's opinions were seemingly retrograde or even reactionary if judged solely by the standards of mid-century American liberalism. Allan Pred's (1983, pp. 92-93) recollections of his encounters with Sauer in the early 1960s point out how this clash of ideologies could create a lack of comprehension and confusion. Pred was disturbed by Sauer's view of the black civil rights movement, which Pred paraphrased as, "Negroes are simple, happy folk whose natural place is close to the soil. If only they hadn't been driven from the southern countryside into the cities we would have none of these problems now." Adjusting for the rhetorical oversimplification implicit in Pred's recounting, this seems to be an accurate rendering of Sauer's probable position. It should be interpreted, however, as a cultural rather than a political conservative position. To some ears it would have sounded like a racist bromide, expecially in the context of the times. Sauer meant it, no doubt, as an anti-modernist's lament for what had been lost. His record of advocacy for the rights of rural, indigenous, and tribal peoples to defend and extend their cultural heritage in the face of "development" and modernization is unambiguous. *Contra* Marx, Sauer championed the "idiocy of rural life" in the sense of *idios* meaning "one's own, separate, distinct." This sentiment and similar ones find company in the persistent "organicist" populist critique of American civilization that has taken different tacks at different times. As Lears suggests, "In our time, the most profound radicalism is often the most profound conservatism" (1981, p. xviii). Sauer probably would have agreed.

Literature Cited

Adams, H. "The Tendency of History." [1894] In *The Degradation of the Democratic Dogma*. New York: Macmillan Co., 1919.

Adams, H. *The Education of Henry Adams*. New York: Houghton Mifflin Co., 1918.

Adams, H. *The Degradation of the Democratic Dogma*. New York: Macmillan Co., 1919.

Billinge, M., D. Gregory, and R. Martin, eds. *Recollections of a Revolution: Geography as Spatial Science*. New York: St. Martin's Press, 1983.

Coffin, A. B. *Robinson Jeffers: Poet of Inhumanism*. Madison: University of Wisconsin Press, 1971.

Contosta, D. R. *Henry Adams and the American Experiment*. Boston: Little, Brown and Company, 1980.

Dicken, S. "The Big Barrens: A Study in the Kentucky Karst." Unpublished Ph.D. Dissertation. Berkeley: Department of Geography, University of California, 1931.

Dunbar, G. S. *Geography in the University of California (Berkeley and Los Angeles): 1868-1941.* Los Angeles: Privately printed, 1981.

Entrikin, J. N. "Carl O. Sauer, Philosopher in Spite of Himself." *Geographical Review,* 74 (1984): 387-408.

Flader, S. *Thinking Like A Mountain.* Columbia: University of Missouri Press, 1974.

Fussel, P. *Abroad, British Literary Traveling Between the Wars.* New York: Oxford University Press, 1980.

Gade, D. W. "The French Riviera as Elitist Space." *Journal of Cultural Geography,* 3 (1982): 19-28.

Geddes, P. *Cities in Evolution.* London: Williams and Norgate, 1915.

Giddens, A. "Modernism and Post-Modernism." *New German Critique,* no. 22 (1981): 15-18.

Goethe, J. W. v. *Italian Journey, 1786-1788.* Translated by W. H. Auden and Elizabeth Mayer. London: Collins, 1962.

Habermas, J. "Modernity versus Postmodernity." *New German Critique,* no. 22 (1981): 3-14.

Hewes, L. "Geography of the Cherokee Country of Oklahoma." Unpublished Ph.D. Dissertation. Berkeley: Department of Geography, University of California, 1940.

Hewes, L. "Carl Sauer: A Personal View." *Journal of Geography,* 82 (1983): 140-147.

Hooson, D. "Carl O. Sauer." In *The Origins of Academic Geography in the United States,* edited by B. W. Blouet, pp. 165-174. Hamden, Connecticut: Archon Books, 1981.

Huizinga, J. *Homo Ludens: A Study of the Play Element in Culture.* Boston: Beacon Press, 1950.

Humboldt, A. v. *Personal Narrative of Travels to the Equinoctial Regions of America.* London: Bohn's Scientific Library, 1852.

Huxley, A. *Beyond the Mexique Bay.* New York: Harper Brothers, 1934.

James, P. E. "The University of Michigan Field Station at Mill Springs, Kentucky, and Field Studies in American Geography." In *The Evolution of Geographic Thought in America: A Kentucky Root,* edited by W. A. Bladen and P. P. Karan, pp. 59-85. Dubuque: Kendall/Hunt, 1983.

Johnston, R. J. *Geography and Geographers: Anglo-American Human Geography Since 1945.* London: Edward Arnold, 1979.

Kenzer, M. S. "Like Father, Like Son: William Albert and Carl Ortwin Sauer." A paper presented at the 81st Annual Meeting of the Association of American Geographers. Detroit, 1985a.

Kenzer, M. S. "Milieu and the 'Intellectual Landscape': Carl O. Sauer's Undergraduate Heritage." *Annals,* Association of American Geographers, 75 (1985b): 258-270.

Kersten, E. W. "Sauer and 'Geographic Influences'." *Yearbook,* Association of Pacific Coast Geographers, 44 (1982): 47-73.

Kniffen, F. B. "Louisiana House Types." *Annals,* Association of American Geographers, 26 (1936): 179-193.

Kramer, J. "Letter from Europe." *The New Yorker* (November 26, 1984): 122-137.

Lears, T. J. J. *No Place of Grace: Antimodernism and the Transformation of American Culture, 1880-1920.* New York: Pantheon Books, 1981.

Lawrence, D. H. *The Plumed Serpent.* New York: Alfred A. Knopf, 1926.

Lawrence, D. H. *Mornings in Mexico.* New York: Alfred A. Knopf, 1928.

Leighly, J. B., editor. *Matthew Fontaine Maury: The Physical Geography of the Sea and Its Meteorology.* Cambridge: Belknap Press, 1963.

Leighly, J. B. "Carl O. Sauer, 1889-1975." *Annals,* Association of American Geographers, 66 (1976): 337-348.

Leighly, J. B. "Scholar and Colleague: Homage to Carl Sauer." *Yearbook,* Association of Pacific Coast Geographers, 40 (1978): 117-133.

Lodge, D. "Modernism, Antimodernism, and Postmodernism." *The New Review,* 4, 38 (1977): 39-44.

Lyotard, J. F. *The Postmodern Condition: A Report on Knowledge.* Minneapolis: The University of Minnesota Press, 1984.

Martin, G. J. "Paradigm Change: A History of Geography in the United States, 1892-1925." *National Geographic Research,* 1 (1985): 217-235.

Mumford, L. *The Story of Utopias.* New York: The Viking Press, 1922.

Mumford, L. *Technics and Civilization.* New York: Harcourt, Brace, 1934.

Mumford, L. *The City in History.* New York: Harcourt, Brace & World, 1961.

O'Brien, M. *The Idea of the American South - 1920-1941.* Baltimore: The Johns Hopkins University Press, 1979.

Parsons, J. J. "Carl Ortwin Sauer, 1889-1975." *Geographical Review,* 66 (1976): 83-89.

Parsons, J. J. "The Later Sauer Years." *Annals,* Association of American Geographers, 69 (1979): 9-15.

Popock, J. G. A. *The Machiavellian Moment, Florentine Political Thought and the Atlantic Republican Tradition.* Princeton: Princeton University Press, 1975.

Post, L. "Cultural Geography of the Prairies of Southwest Louisiana." Unpublished Ph.D. Dissertation. Berkeley: Department of Geography, University of California, 1937.

Pred, A. "From Here and Now to There and Then: Some Notes on Diffusions, Defusions and Disillusions." In *Recollections of a Revolution: Geography as Spatial Science,* edited by M. Billinge, D. Gregory, and R. Martin, pp. 86-95. New York: St. Martin's Press, 1983.

Price, E. "Mixed Blood Racial Islands of Eastern United States as to Origin, Localization, and Persistence." Unpublished Ph.D. Dissertation. Berkeley: Department of Geography, University of California, 1950.

Sauer, C. O. *The Geography of the Ozark Highland of Missouri.* Bulletin No. 7. The Geographic Society of Chicago. Chicago: University of Chicago Press, 1920a.

Sauer, C. O. "The Economic Problem of the Ozark Highland." *Scientific Monthly,* 11 (1920b): 215-227.

Sauer, C. O. " The Morphology of Landscape." *University of California Publications in Geography,* vol. 2, no. 2 (1925): 19-53.

Sauer, C. O. *Geography of the Pennyroyal.* Series 6, vol. 25. Frankfort: Kentucky Geological Survey, 1927a.

Sauer, C. O. "Recent Developments in Cultural Geography." In *Recent Developments in the Social Sciences,* edited by E. C. Hayes, pp. 154-212. Philadelphia and London: J. P. Lippincott, 1927b.

Sauer, C. O. to H. W. Odum, January 23, 1935. *Howard W. Odum Papers,* Southern Historical Collection. University of North Carolina, Chapel Hill, 1935a.

Sauer, C. O. to H. W. Odum, February 7, 1935. *Howard W. Odum Papers,* Southern Historical Collection. University of North Carolina, Chapel Hill, 1935b.

Sauer, C. O. "American Agricultural Origins: A Consideration of Nature and Culture." In *Essays in Anthropology Presented to A. L. Kroeber in Celebration of His Sixtieth Birthday, June 11, 1936,* pp. 278-297. Berkeley: University of California Press, 1936.

Sauer, C. O. "Destructive Exploitation in Modern Colonial Expansion." *Comptes Rendus du Congrès International de Géographie, Amsterdam.* Vol. 2, Section 3c (1938a): 494-499.

Sauer, C. O. "Theme of Plant and Animal Destruction in Economic History," *Journal of Farm Economics,* vol. 20 (1938b): 765-775.

Sauer, C. O. "Foreword to Historical Geography," *Annals,* Association of American Geographers, vol. 31 (1941): 1-24.

Sauer, C. O. *Agricultural Origins and Dispersals.* Bowman Memorial Lectures. Series 2. New York: American Geographical Society, 1952.

Sauer, C. O. "The Education of a Geographer." *Annals,* Association of American Geographers, vol. 46 (1956): 287-299.

Sauer, C. O. "Status and Change in the Rural Midwest—A Retrospect." *Mitteilungen der Öesterreichischen Geographischen Gesellshaft,* vol. 105 (1963): 357-365.

Sauer, C. O. *Northern Mists.* Berkeley: University of California Press, 1968.

Sauer, C. O. "Casual Remarks." *Historical Geography Newsletter,* 6 (1976): 69-76.

Sauer, C. O. to J. H. Willits, January 1, 1942. In *Andean Reflections,* edited by R. C. West, pp. 9-11. Dellplain Latin American Studies, no. 11. Boulder, Colorado: Westview Press, 1982.

Smith, H. N. *Virgin Land: The American West as Symbol and Myth.* Cambridge: Harvard University Press, 1950.

Speth, W. M. "Carl Ortwin Sauer on Destructive Exploitation." *Biological Conservation,* 11 (1977): 145-160.

Speth, W. W. "The Anthropogeographic Theory of Franz Boas." *Anthropos,* 73 (1978): 1-31.

Speth, W. W. "Berkeley Geography, 1923-33." In *The Origins of Academic Geography in the United States,* edited by B. W. Blouet, pp. 221-244. Hamden, Connecticut: Archon Books, 1981.

Stevenson, E. *A Henry Adams Reader.* Garden City: Doubleday and Co., 1958.

Thomas, W. L., Jr., editor. *Man's Role in Changing the Face of the Earth.* Chicago: University of Chicago Press, 1956.

Thornthwaite, C. W. "Louisville, Kentucky: A Study in Urban Geography." Unpublished Ph.D. Dissertation. Berkeley: Department of Geography, University of California, 1930.

Trindell, R. T. "Franz Boas and American Geography," *The Professional Geographer,* 21 (1969): 328-332.

Weintraub, K. J. *Visions of Culture.* Chicago: University of Chicago Press, 1966.

West, R. C. *Carl Sauer's Fieldwork in Latin America.* Dellplain Latin American Studies, no. 3. Ann Arbor: University Microfilms International, 1979.

Williams, M. "'The Apple of My Eye': Carl Sauer and Historical Geography." *Journal of Historical Geography,* 9 (1983): 1-28.

Wright, J. K. *The Geographical Lore of the Time of the Crusades.* New York: American Geographical Society, 1925.

Wright, J. K. "On Medievalism and Watersheds in the History of American Geography." In *Human Nature in Geography,* by J. K. Wright, pp. 154-167. Cambridge: Harvard University Press, 1966.

Zelinsky, W. "Settlement Patterns of Georgia." Unpublished Ph.D. Dissertation. Berkeley: Department of Geography, University of California, 1953.

Photographs

CARL ORTWIN SAUER, A.B., B.S., Warrenton, Mo.

Pulse Staff, Epworth League Cabinet, Secretary Class, Goethenian, Assistant Teacher (French).

"My native village produced at least one great man."

P.V. Somewhat of a selachian in social circles, also a walking encyclopedia of professional base ball.

P.F. Those peg top trousers.

Carl Sauer's yearbook photograph. "P.V." = *principal virtue; "P.F."* = *principal fault.*
Source: *'08 Pulse [C.W.C.],* edited by P. H. Walter. Warrenton, Missouri: Central Wesleyan College, 1908. Photo courtesy of Margaret Schowengerdt.

Carl Sauer with wife, Lorena (Schowengerdt) Sauer, and son, Jonathan.
Photo circa 1919, courtesy of Margaret Schowengerdt.

Carl Sauer in the early or mid-1930s.
From the Berkeley Geography Department Collection. Photo courtesy of Henry J. Bruman.

Only a plea that we needed a scale would persuade Sauer to pose with this rare bonete, a relative of the papaya.
Photo taken in May 1939 by Henry Bruman, courtesy of Henry J. Bruman.

A second bonete variety got only a view from the rear by Sauer.
Photo taken in May 1939 by Henry Bruman, courtesy of Henry J. Bruman.

117

Sauer in the field (Tecomán, Colima)
Photo is from 1939, courtesy of Henry J. Bruman.

Sauer relaxing on a trip to Mexico with Mrs. Sauer and daughter, Elizabeth, in 1941.
From the Berkeley Geography Department Collection. Photo courtesy of Henry J. Bruman.

118

Sauer in 1941 on a trip to Mexico.
From the Berkeley Geography Department Collection. Photo courtesy of Henry J. Bruman.

119

A typical Sauer mood in office or seminar.
Photo taken by Dorothea Lange, 1944. Photo courtesy of Henry J. Bruman.

Carl Sauer and Mrs. Sauer in Berlin in 1959 on the occasion of the awarding of the Alexander von Humboldt Gold Medal of the Gesellschaft für Erdkunde zu Berlin.
Photo taken by Chauncy Harris, courtesy of Chauncy D. Harris.

Carl Sauer on Arch Street in Berkeley, California.
Photo taken by Hugh Iltis in 1966. Photo courtesy of William M. Denevan.

H. L. Sawatzky (left) and Sauer, taken in October 1968 in Winnipeg, Manitoba.
Photo taken by Lorena Sauer. Photo courtesy of H. L. Sawatzky.

Carl O. Sauer in his office, April 1970.
Photo taken by Berl Golomb. Photo courtesy of Herbert M. Eder.

PART THREE

Students' Perspectives

Carl Sauer in Midcareer:
A Personal View
by One of his Students

Henry J. Bruman

In appearance Carl Sauer was not remarkable. He was a man of middle height, somewhat stocky, slightly stooped in later years. He matured early and usually looked older than his age. As time went on his face became deeply lined, a legacy not only of years of fieldwork under the desert sun of the American Southwest and northern Mexico, but also of recurring bouts of gastrointestinal illness caused by contaminated food in the field. His gaze was direct, and his eyes were perhaps the most notable feature of his appearance. But it was when he spoke that one really began to take notice. It took only a minute of conversation to make one realize that here was an intellect!

My first meeting with Sauer was unforgettable. It introduced me to the most powerful personality I have ever known. His influence on me and my cohorts at the University of California, Berkeley, was profound and pervasive and has remained with many of us through years of travail, through occasional feelings of inadequacy in the face of Sauer's brilliance, and even through rare disputes with Sauer himself. It must be realized, of course, that these recollections are filtered through almost fifty years of memory. During that half-century my memories have not become any clearer, even though my convictions and conclusions about Sauer may have firmed up. But first let me digress briefly to introduce myself as I was fifty years ago.

I already knew that I wanted to become a university professor. That decision had been reached owing to the influence and friendship of my first great mentor, Frederick C. Leonard (1896-1960), Professor of Astronomy at the University of California, Los Angeles. I had an A.B. in chemistry from U.C.L.A. but felt an increasing inadequacy in myself as a potential chemist. I was simply not a good enough laboratory technician, and had twice already ruined organic preparations in their final stages of synthesis, after some two weeks of effort, by dropping them on the floor. It was frustrating and I finally got the message. At about the same time a different influence entered the picture. I had met a young lady who shared my interest in the outdoors and in mountain climbing, and in the winter of 1933-34, when I was twenty, we decided to try the as yet uncompleted Pan-American Highway to Mexico City. She and I and her mother and a young faculty member from the Women's Physical Education Department at U.C.L.A. drove down as far as Taxco and the Caves of Cacahuamilpa, spending six weeks and camping out all the way. We were deeply impressed by what we saw. The next summer I decided to go to Mexico again, alone, and take work in the *Escuela de Verano* of the National University. One of the courses I took was on the geography of Mexico, given in Spanish by Professor Luís Osorio Mondragón (1883?-1944). It was the fascination of this course that made me switch to geography.

In 1934-35, my last year as an undergraduate, I began to take geography courses at U.C.L.A. while completing the chemistry major, and the following year I was an unclassified student concentrating on geography. When the time came to decide where I should do graduate work, Myrta L. McClellan (1875-1963) and Ruth E. Baugh (1889-1973), two of my teachers, suggested that, everything considered, Berkeley was probably the best place, and that I should try to get accepted by Professor Sauer, even though they were personally not quite sure that he was a gentleman.

After suitable arrangements we met in Mr. Sauer's large basement office in Agriculture Hall. My first impression was that of a kindly, somewhat portly, middle-aged man with an expressive, slightly wrinkled face, comfortably dressed in an old suit, and smoking a pipe. He began by asking me about my background. I told him about the chemistry, my minor in astronomy, my two trips to Mexico, including a failed attempt to climb Popocatépetl and a horseback trip to the Totonac Indians around Papantla to study their vanilla industry. He seemed to approve. I then told him that I had come to America as an immigrant from Germany in 1922, aged nine, with my mother and stepfather. That interested him. "So, you speak German. That will be very useful." He asked about other languages, and I told him I could handle Spanish and could read French. "Fine." Then he started on a discourse on something related to Indians in the State of Vera Cruz that had been brought to mind by my trip to the Totonac. He talked slowly and well, apparently feeling his way and obviously thinking out his phrases as he went along. I had never seen anyone put so much thought into a conversation *before* the words were uttered; and I never have since. The sentences rolled out in beautiful cadences practically ready to print. This remains one of the strongest impressions that I retain of Carl Sauer to this day; this and the related characteristic of his phenomenal memory that could retain and, when needed, recall a host of other facts bearing on the subject under discussion.

But back to the first interview. Every so often Sauer would pause to light his pipe, which I interpreted as a signal to ask a question or interject a comment based on my own background. He would respond to these in his fashion, replying more fully and further developing a theme when he considered my contribution of interest, and ignoring interjections that were of little or no value. After this had gone on for several rounds, he stopped dead. He lit his pipe again, shifted his weight, and got out his pocket knife to start splitting one of the large kitchen matches that he commonly used, but he said nothing. Now and then he looked at me. I could feel myself starting to sweat. "This is it," I thought, "either you come through now, or you're out." So, in desperation, I reached down to levels I had not plumbed before, and started talking of a problem I had encountered with the Totonacs and their culture of the vanilla orchid. I asked, "How could this group of Indians have learned to give the vanilla flower the necessary pollination by hand, without which there is practically no production of vanilla pods? This implies a biological sophistication most unusual among American Indians."[1] I described the process in detail and pointed out that for much of the tribe the sale of vanilla pods was the main source of income. And then I stopped. I was wrung out and could find nothing more to say. Sauer continued to sit and look and I was afraid I had flunked the interview. After what seemed a very long time but was probably only a few seconds, he said, "Yes, Bruman, but have you thought of this?" And then he was off on another of his rounded periods, and I could relax. We kept on talking for more than an hour. When I got up to go he saw me to the door and said, "Yes, come on up for the fall semester. We'll find a place for you."

And so in the summer of 1936 I joined the group of teaching assistants and other graduate students whom Jan Broek (1904-1974), who was there at the time as Assistant Professor, dubbed the B-hive: Bowen, Bowman, Bruman, and Blumenstock, and of course Broek himself, all working with or under the Doc, as we called him—but never to his face. There were still others,[2] of course, but the preponderance of Bs and the alliteration of the names were remarkable. Our duties as teaching assistants were those of section leaders and readers of exercises and examinations. In the sections we learned how to give lectures and after hours we practiced preparing exam questions. As a rule we audited all the courses where we helped out, and of course took advanced courses and seminars of our own, including Sauer's Saturday field course.

It was in the lecture that Sauer's superb gifts with words and ideas became evident to all of us. I had an initial glimpse in my first interview, but now it became a daily feast. He spoke slowly, choosing his words with care; you could practically see him thinking before your eyes. His finest lecture courses were his lower-division course on cultural geography and the two upper-division regionals on North America and Latin America. He had an extraordinary gift for colorful language, and we all felt that we were witnessing the process of literary creation in action. The students were fascinated, even enthralled, by the intellectual activity taking place before them. I recorded in my notes several examples of this sort. Once in the Latin America course he characterized the volcanoes of Central America as "the external

127

manifestation of the suppuration of a tortured terrestrial epidermis.'' This simile is perhaps not entirely accurate from a geophysical point of view, and it might be faulted as excessively and unnecessarily colorful, but its very vividness lent an extraordinary graphic immediacy and made the images indelible. It was an excellent and highly successful teaching device. You could sense the stirring and the murmurs of appreciation on such occasions. Twice in my presence students broke out in spontaneous applause after one of these descriptions. At these demonstrations Mr. Sauer would look up, momentarily startled, smile slightly, and go on. Another time, talking about house types on the Mexican plateau, he remarked that as a rule buildings were put together of worked stone "except for those lesser whitewashed structures in which the Mexican population perpetuates itself." And once, in the elementary cultural geography course, he ascribed the temporary eclipse of French military strength and the slowdown of economic and political growth in the nineteenth century to the fact that "Napoleon fertilized the soils of Europe with the best young blood of France." This may be a quote from some other source, but I have no evidence for that. In either case it is a good example of Sauer's occasionally very colorful lecture style and his predilection for the startling figure of speech.[3]

Sauer felt no compunction to be comprehensive or systematic in his course presentations. The fact that he did not cover a region systematically in the manner of standard geography courses elsewhere did not bother him. He felt it more useful and more interesting to emphasize problem areas and areas of his own research interests. His attitude toward his lecture courses was more like that of a German university professor: the missing parts were meant to be filled in by the students through individual investigation and reading. Thus in his New World regional lecture courses he would be strong on physical geography, plant geography, aboriginal culture, domestication of plants and animals, early European contacts, the early phases of colonial economies, and similar topics on which he had published or was actively doing research, and he would usually ignore the modern scene. In the mid-1930s he loved to spend large sections of time on plant geography—he kept a first edition of Humboldt's *Essai sur la Géographie des Plantes* (1805) in the office—and especially on the origin and spread of domesticates. When Joe Spencer (1907-1984) took the North America course in the early thirties the last lecture had American settlers just peeking over the Appalachians in their colonizing trek westward. When I took it, in 1937, we ended up in Newfoundland, having spent only an hour on the colonization of the Atlantic seaboard farther south.

A regional course to Sauer was for presenting and analyzing logical and historical interrelationships between the land and its human inhabitants. It was never merely to "cover" an area with a succession of facts. The core of his regional interests lay in the determination and exploration of cultural origins, in the dispersal of cultures and their traits through time and space, and their growth, modification, and extinction. A second major concern was with human use and misuse of the land. These were his major research themes, and they tended to find expression again and again in his conversation, his lectures, and his writings. He left to others such fields as political geography, urban geography, and modern economic geography, sometimes expressing a personal lack of interest in them, occasionally commenting acidly on current publications in these subdisciplines, especially when he felt the writers displayed no real sense of problem or ignored the importance of the time factor.

Carl Sauer did not engage in small talk. To him a conversation had to have a purpose, at least an implied purpose, and it had to have a respectable content. He had what the writer Joseph Epstein has called "gravity," which "derives from a serious . . . mind, unencumbered by the clichés of the day, at work on serious matters" (Epstein 1985, pp. 342, 96). A talk with Sauer was always instructive, sometimes provocative, occasionally memorable. He was usually at his best before a handful of interested, knowledgeable students like a seminar group or a field party. In the presence of such an attentive audience he would relax, sometimes even lean back and put up his feet, and always lighting, smoking, and relighting his pipe. Often he would go to work on some of his spent matches, whittling, or rather splitting them with his pocket knife into halves, quarters, and slivers, because eighths were not really possible. Thus, completely at ease and with his hands pleasantly busy he would give his mind free rein to retrieve and analyze items from his memory and correlate them with the subject at hand. He was superb at it. Not that he was always right. Sometimes he let his enthusiasm run away with him, and sometimes he hung on to his notions too long when the contrary evidence was stronger, but these quirks just show that he was human; it was good for the rest of us to see that the master also had a few limitations.

Conversation to him was often a means of ordering and correlating his own store of knowledge in advance of writing up his research results, and when he actually got down to writing the process seemed to flow along without too much difficulty. In these conversations he would sometimes get carried away and forget to pay attention to other matters. I remember once he had been holding forth before a small group for the better part of an hour, talking constantly except for the occasional chiming in of a student. We were all under his spell, captivated not only by his phenomenal memory and intellectual power, but also by the pile of match shavings that was mounting higher and higher and the angle of his chair, which was getting lower and lower. Finally gravity triumphed, and he went down. When it became apparent that the only damage was to his dignity we could not help laughing, and he joined in. "Must remember to get that chair fixed," he said, and went on with his discourse, hardly missing a beat.

The Saturday field course in the mid-1930s would usually concentrate on different portions of the east bay region of San Francisco, although one semester we went to Marin County. We would go out in a caravan of three or four cars, Mr. Sauer in the lead. It was an honor to be asked to ride in his car, especially in the front seat, and the occupants were changed frequently. One car was usually a university car, and the rest belonged to students or junior faculty. We all chipped in on expenses, and we brought our lunch in a brown bag.

When Mr. Sauer stopped we would do the same and pile out. I remember innumerable encounters with barbed wire fences which we usually negotiated by crawling under or between the strands, in time becoming quite expert. There were many traverses of sloping pastures, and there was much studying of the effects of erosion: slumping, sheet wash, rilling and gullying. The function of cattle tracks and gopher holes in the erosion of surfaces of varying slope was analyzed. For our lunch break we would usually contrive to be on a hilltop somewhere, preferably in the shade, with a view of the landscape. After lunch Mr. Sauer would talk, perhaps about the morning's observations, but more often on other relevant topics, such as Walther Penck's (1888-1923) ideas of landscape development, which he thought useful, and we would ask questions. At times he would discuss the view before us, pointing out accordant summits, remnant terraces, recurring slope changes, evidence of tectonic displacement, the distribution of vegetation associations in relation to slope and exposure, and similar topics. Later in the semester one or more individual students might be asked to interpret another landscape for the group. Thus we all got practice in observation and field analysis. Settlement geography was not neglected; every so often we found evidence of earlier Indian habitation, where shell mounds and buckeye stands were evaluated. But we paid little attention to modern settlement, agriculture, transportation and the like.

A term paper was required of those taking the course for credit. Once I wrote on the genesis and preservation of cattle terraces on hillsides and got an "A." Another time I did a much more laborious job of field mapping crops in part of Contra Costa County and then correlating the field map with standard maps of soil and slope. But because Sauer considered the paper to be mainly a project, rather than being problem oriented, he graded it only a "B."

The longer field trips to the Southwest and to various parts of Latin America have already been well summarized by Bob West, with comments by numerous participants (West 1979). I was in the field with Sauer and Isabel Kelly (1906-1982) in Jalisco, Colima, and Michoacán during most of May and June, 1939, after working in Guatemala, Honduras, and El Salvador, finishing up a year's field research that led to the dissertation. It was good to see Mr. Sauer again, and I gained some new perspectives from the experience. One of my strongest recollections is of the remarkable empathy he was able to establish with field informants. He was genuinely interested in them and they took to him. His colloquial Spanish and his unpretentious, avuncular manner made for easy rapport. He had a way of simply going down to whatever gathering place there was, a bench in the plaza, for instance, and in a short time finding himself in active and fruitful conversation in the middle of a group of locals. He took no written notes, since they could have curtailed the spontaneous exchange, but he sometimes made records later. These informal group encounters were so frequent that they became an integral part of his field technique. There was a bit of trouble only once, as I recall, somewhere in the western lowlands in the border country between Colima and Michoacán. In a village out of which we had worked for several days the local priest got it into his head that Sauer was spreading antireligious or anticlerical propaganda and that we were in league with the devil. After a sermon against us the attitude of the local people became unfriendly and we left.

In the field one gets to know one's companions, and a few additional vignettes of Sauer come to mind. Although not overtly religious, he once expressed to me his appreciation of the strong Christian faith of the German farmers of Missouri, as well as of their Lutheran hymns. He did add, in a characteristic aside, that nothing is quite so hard on the ears as the singing of a Protestant church choir made up of Missouri farm folk.

Every so often, even in the field, Mr. Sauer would revert to being the professor. I had visited the Huichol Indians in their *barrancas* in December 1938 to gather data about their allegedly native beverage still, and here are some comments about them which Mr. Sauer made to me on the spur of the moment in a conversation one evening, as recorded in my notebook:

The Doc's ideas about the Huicholes—June 1939

(1) They are a western branch of the Guachichiles.

(2) The Tecual, at first called Guachichiles, were once living just north of Tequila. They were probably the southernmost Huicholes.

(3) Orozco y Berra's statement that the Guachichiles did not speak a Nahua language is probably without foundation.

(4) The Guachichiles were not agriculturalists. The western branch (Huichol) learned agriculture probably through contact with their more advanced neighbors.

(5) The low cultural standard of the ancestors of the Huicholes is even today seen in the lack of emphasis on pottery making.

(6) The Cora have always looked down on the Huicholes as an inferior group.

(7) The Cora themselves probably degenerated from a higher cultural level.

This was a very impressive performance, compounded as it is of obscure facts based on wide reading, and on personal reflection and judgment. And it was impromptu.

Another memorandum in the same notebook records some of Sauer's reasons for favoring a South American origin for the coconut instead of a Southeast Asian one. The present consensus is that he was in error, and I will come back to this topic in just a moment. Let me say, however, that even when he was wrong or out on a limb he could be very persuasive, and he rarely admitted that he had been mistaken. His way was to change the emphasis or go on to something else. At a later time he did not always feel bound to defend prior positions. A real waste of time in his view was methodological dispute of the "What is geography?" sort. He simply refused to engage in it and once remarked, "Most of my eastern colleagues seem to think that all I do here in Berkeley is sit in my office and read the 'Morphology of Landscape.'" He also considered personal vendettas as worse than useless and refused to participate. When E. D. Merrill made his attack on Sauer's competence as a biogeographer, specifically with respect to the facts and generalizations in his *Agricultural Origins and Dispersals* (Merrill 1954, pp. 271, 275-6, 279, 282, 285, 287, 345), Sauer refused to reply. One point of difference was Sauer's inclination to accept O. F. Cook's argument that the genus *Cocos* was native to South America because most of the closely related genera were native there, overlooking or undervaluing other powerful evidence that points to an origin in Southeast Asia or adjacent islands (Cook 1901, 1910). It may be that Merrill in his old age transferred some of his animosity toward Cook to Sauer, for the printed attack goes well beyond acceptable limits of scholarship. Judging by some comments Merrill makes in passing (1954, pp. 175 and 299) the editors actually toned down some of his language and kept the text from being even more intemperate.

I personally had a long dispute with Sauer about the problem of the coconut's origin. I had worked on the early history of the coconut in the Americas and its connection with the Colima coconut wine industry, the introduction of an essentially Filipino still into Huichol culture, the Manila Galleon, and the coming of Filipinos to Mexico. Some of this material was included in my dissertation, some of it was worked on and published subsequently. As a result of this work I became convinced that, even if it were true that the coconut had most of its wild relatives in this hemisphere, *Cocos* as a genus and cultigen is a product of the other side of the Pacific. Its antiquity in the Eastern Hemisphere goes back many thousands of years, as does its use by humans. In aboriginal America it was not a staple among any native groups; in fact it occurred only at a few separated locations along the Pacific Coast where a casual and relatively recent introduction from across the Pacific seems most likely.

My articles on the coconut appeared during the forties (Bruman 1944, 1945, 1947). Sauer's article on South American useful plants for volume six of the *Handbook of South American Indians* appeared in 1950 (Sauer 1950), although I had seen a copy of the manuscript at the Smithsonian as early as 1944. After I had read the manuscript I congratulated him on it, but told him I could not agree with his stand on the coconut and hoped he would reconsider before publication. That began an interchange of letters on the subject which is now part of the Sauer Papers at the Bancroft Library.[4] I could not budge him and it hurt. I had not encountered this trait in him before, although two or three of his other students have mentioned somewhat similar episodes in their comments on the field trips as recorded by West (see West 1979).

The unfortunate result of this exchange was that it created a gulf between us that took more than a decade to bridge over. Being a Sauer student was not always an unmitigated blessing. He was at times almost too impressive, and for some of us his dominance became almost overpowering. It took me a long time to reconcile myself to the thought that the Doc and I could not agree on an interpretation of the facts regarding a basic research problem. These feelings of doubt and inadequacy on my part were exacerbated in 1948 when I gave a talk at Stanford University at an Americanists meeting on the role of geography in area studies. Mr. Sauer was in the audience, at the rear of the room. After the conclusion of the paper and the subsequent discussion, the audience filed out through the door, which happened to be near the front. I was still at the podium. When Sauer left he walked right past me, without a word, without so much as looking at me, although I was moving toward him and he must have sensed that I wanted to greet him after not having seen him for several years. But no, I had crossed him, and what I was then working on was not to his liking. He could be a hard man.[5]

Fortunately, with the passing of the years he gradually mellowed. When he gave a seminar at U.C.L.A. in 1961 (see Sauer 1963) it was almost like old times. But he could still be difficult. He is the only person ever who, on visiting me in my office at U.C.L.A., has automatically sat down on my chair at my desk in my presence. Not many of us would do this in another person's office, even if he or she were a former student. After all, we each have our feeling of territoriality. But I did not raise a fuss. There was something rather reassuring about the gesture. The Doc was back.

Let me close with an anecdote from even further back, which shows that even as a graduate student Sauer exhibited the extraordinary mental power which became the hallmark of his career. On one of his visits to Berkeley, about 1936, I heard Bob Hall (1896-1975) of the University of Michigan tell a story that illustrates as well as anything I know the quality of Sauer's remarkable mind. It seems that Sauer's fellow graduate students at Chicago some twenty years earlier got sick and tired of his know-it-all manner and decided to show him up. What was especially galling was that Sauer was not bluffing: he actually knew what he was talking about. It is not easy to forgive a fellow who knows all the answers. But it is almost impossible to forgive him when he usually turns out to be right. So they decided to gang up on him. They would read up on some geographic topic in advance, each taking a particular aspect, and they would spring their knowledge on Sauer someday when they were all together and so show him up and teach him a lesson.

After some discussion they decided the topic should be cheese: regional varieties of cheese in different parts of the world, including sources of milk, methods of manufacture and marketing, consistency and flavor of finished product, and so forth. So they divided up the topic, each did his separate research, and the moment came a couple of weeks later to display their new knowledge in Sauer's presence. One man casually started talking about some reading he just happened to have done about cheddar cheese, about the differences among New York, Wisconsin, and Oregon varieties, and how they all related to earlier forms of cheddar in England. Then another chimed in, adding what he just happened to know about other English cheeses, mentioning the Derby, the Cheshire, the Lancashire, and then the Stilton, the noblest of them all, which is aged longer than the others and owes its magnificent flavor in part to the development of a green mold. Then the third student began with a discussion of mold cheeses in the continent of Europe, blue cheese in Denmark, Roquefort in France, Gorgonzola in Italy. Then the fourth gave his report, and the whole ostensibly impromptu performance went on for a good half hour. Bob Hall remarked that by the time the first talk on cheddar was over and the second one on English cheeses began ''Sauer's moustache began to broaden.'' He could smell rats in the cheese.

And then, when the group was quite through displaying their new-found expertise, Sauer began. With no preparation he talked in his careful, measured way about cheese making among the German settlers in Missouri, and how it differed from the practices of settlers in Pennsylvania and Wisconsin. Then he went into cheese making in Bavaria, Switzerland, and Austria, and told them all in all far more than they had discovered in their combined investigations. Well, that was it. The others were licked and they knew it. They never baited him again.

This little story illustrates Carl Sauer's wide knowledge, retentive memory, and overall precocity at an early stage in his professional career. The personal traits of industry, superb recall, and powerful intellect stayed with him to the end.

In the world of learning we need people who ponder about fundamental issues, about values and ends; just as we also need people who become teachers and administrators and thus keep the machinery going; just as we also need people who acquire wealth and to a degree filter it back through charitable gifts to promote the public good. But among these the noblest group is certainly the first, which charts new horizons, identifies new issues, finds breakthroughs for old problems, and advances human learning and culture. This is where Carl Sauer labored productively, growing and changing and learning new things to the end of his life. As Goethe said, that which is becoming, not that which has become, contains the divine spark.

Notes

1. I finally found the answer years later (cf. Bruman 1948).

2. One of the others, Ann Nicholls, came from Australia by way of Canada to spend the academic year 1937-38 at Berkeley. As Mrs. Ann Marshall she has had an active career in Australian universities and has been a recent Lewis Medalist of the Royal Geographical Society of Australasia (South Australian Branch). She has kindly sent me a few of her own recollections to add to mine, and I present some of her comments below:

> I think you are dealing with the most important period of Sauer's teaching. We *knew* we were special. We could recognize one another years later in different parts of the world. There was an intellectual focus in the group and it came from his stimulus. We talked geography over endless cups of coffee after seminars. The source of the glow was the Doc.

> You said once you had to revisit Berkeley now and again to get your batteries recharged. . . . I would never have considered talking to him about anything other than work. He never failed me in three periods at Berkeley if I wanted a searchlight directed on a professional problem. [But] do you remember ordeal by silence? How if you took him a question he would sit and think for minutes—giving you time to go over the question and wonder if it sounded silly? That was a cruel technique. He could have said "I'll have to think about that for a moment." I have never treated a student like that. . . . Did the Doc get more out of us by being remote?

3. Marshall recorded in her notes from the cultural geography course, referring to the early spread of Islam, "The barbarians, lean and tough from hard living, came to plunder and remained to be educated," and "The Arab remained the country boy come to town, not accepting but realizing the value of city ways."

4. As I reread those letters I am impressed with the kindness with which Sauer states his case, and the tentativeness he ascribes to current solutions of the coconut origin problem, including his own, just as I am appalled at the brashness of young Bruman feeling his oats. Thank God we can still learn with age.

5. At U.C.L.A. the Latin Americanists had been doing some thinking about starting an interdisciplinary major in Latin American Studies, and this paper was part of that effort. Sauer for years had loathed the thought of any sort of interdisciplinary activity as being hopelessly mired in committee work, unless it was strictly research oriented, and even then to be approached with skepticism. But to see Bruman, who had already gotten into print against his advice on the origin of the coconut, turn so completely to the currently fashionable, but in his view brainless, mirage of interdisciplinary committee work, may well have seemed to him unforgivable.

Literature Cited

Bruman, H. J. "Some Observations on the Early History of the Coconut in the New World." *Acta Americana*, vol. 2 (1944): 220-243.

Bruman, H. J. "Early Coconut Culture in Western Mexico." *Hispanic American Historical Review*, vol. 25 (1945): 212-223.

Bruman, H. J. "A Further Note on Coconuts in Colima." *Hispanic American Historical Review*, vol. 27 (1947): 572-573.

Bruman, H. J. "The Culture History of Mexican Vanilla." *Hispanic American Historical Review*, vol. 28 (1948): 360-376.

Cook, O. F. "The Origin and Distribution of the Cocoa Palm." *Contributions from the United States National Herbarium,* vol. 7 (1901): 257-293.

Cook, O. F. "History of the Coconut Palm in America." *Contributions from the United States National Herbarium,* vol. 14 (1910): 271-342.

Epstein, J. *Plausible Prejudices: Essays on American Writing.* New York and London: W. W. Norton & Company, 1985.

Humboldt, A. de and A. Bonpland. *Essai sur la Géographie des Plantes.* Rédigé par Al. de Humboldt. Paris: Levrault, Schoell et compagnie, 1805.

Merrill, E. D. "The Botany of Cook's Voyages." *Chronica Botanica* 14: 5/6. Waltham, Massachusetts, 1954.

Sauer, C. O. "Cultivated Plants of South and Central America." In *Handbook of South American Indians,* edited by J. H. Steward, pp. 487-543. Vol. 6 (Smithsonian Institution, Bureau of American Ethnology, Bulletin 143), Washington, D.C.: U.S. Government Printing Office, 1950.

Sauer, C. O. *Plant and Animal Exchanges Between the Old and the New Worlds: Notes from a Seminar Presented by Carl O. Sauer,* edited by Robert M. Newcomb. Los Angeles: Los Angeles State College, 1963.

West, R. C. *Carl Sauer's Fieldwork in Latin America.* Ann Arbor: University Microfilms International, 1979.

Carl Sauer, A Self-Directed Career

Homer Aschmann

My first meeting with Mr. Sauer was in late summer 1941. My dissertation was signed nearly thirteen years later, but I continued to regard him as a mentor the rest of his life. Entering graduate students were graciously granted an interview that was only partly an interrogation and could last one or more hours. They soon learned that further such one-on-one discussions had to be earned by uncovering something new and interesting to Sauer in the field or library. The former was easier than the latter.

Graduate students in geography coming to the University of California, Berkeley, from elsewhere during the 1930s and early 1940s were likely to know of Mr. Sauer from having been assigned "The Morphology of Landscape" (Sauer 1925) as required reading for the serious student. The erudition and grimly rigorous philosophical argumentation it displayed, enlivened only by the "bagging their own decoys" thrust at geographic determinists, gave him a formidable if not awesome image to the new student. It was a pleasant shock to learn that that paper was not in the Berkeley curriculum nor did it significantly inform the department's teaching and investigation. In the "Morphology" Sauer's logical analysis drove him to characterize geography as a generic inquiry. Immediately after writing it his real interests won out, and his investigations became genetic, particularly into the origins of cultural landscapes.

Sauer's monumental knowledge tended to overwhelm most students. On frequent occasions he knew more about your seminar report than you did! Examination questions could be impossibly difficult though he graded them easily, giving credit for any spark of understanding rather than demanding encyclopedic iteration. The new graduate student was initiated by his or her more senior fellows into a subculture that was defensive and mutually helpful. It gradually became apparent that Sauer was really driven by curiosity. Each set of observations aroused questions as to origins and interrelationships, and it was Sauer's teaching style to explore such observations further with his students, especially in seminars. Could the student throw new light on a question derived from random or focused reading or from direct observation? The practice of seeking obscure references in the library and trying to observe

137

both patterns and small, significant details in the landscape on journeys and field excursions was encouraged and rewarded by Sauer's enthusiastic interest when one could add to his knowledge and understanding.

Perhaps 1941 was the last year he formally taught the Field Course (GEO. 101). It was singularly informal with seemingly random stops to view a road cut, a landscape, an abandoned farmstead, or the wild or weedy vegetation. These would provoke questions such as, "Did the soil-profile give evidence of climatic change?" or, "On the basis of surviving plantings around the abandoned farmstead could one determine the home state of its former occupant?" A term report or thesis might grow from following up on one or a set of such questions. I had the fortune to be in the field with Mr. Sauer for a few weeks in Baja California and observed another aspect of his fieldwork. He was extremely effective in engaging *campesinos,* especially in isolated rural areas, in conversation. No planned line of questioning, let alone a list of questions, was used. The effort was to find a topic the individual knew about and wanted to talk about. It was likely to relate to some aspect of the local area.

The quality Mr. Sauer sought in his students—and if they did not have it in some measure they would not continue to study under him—was the ability to become so interested in a question that pursuing it as far as you could was fun. In that way he was seeking to reproduce his own image, but his brilliance and spectacular memory were not really required. On the other hand, he was completely intolerant of bluffing or intellectual dishonesty. The story is told that while he was serving as third reader on a thesis in the History Department that had already been signed by Berkeley's most famous historian, he thought that something looked familiar. He proceeded to the library, found the thesis that had been copied and which had been signed by the same historian a number of years before, and then placed both theses on the professor's desk!

The Berkeley ambience had a special quality that we were encouraged to take advantage of. Somewhere in the community there was likely to be a specialist on the most esoteric subject you could become concerned with, often in an unexpected place. It was the culture of the community—one in which Sauer participated fully with students from all over the campus—that if the student had learned enough about the topic to discuss it effectively that discussion was freely available. On my dissertation the greatest aid came from Sherburne F. Cook (1896-1974), a physiologist affiliated with the medical school, and for a seminar report on beer I could consult with Emil Mrak (1901-) in the Food Technology Department; Mrak was soon to become the notably successful Chancellor at the University of California, Davis.

Qualifying for the doctorate in geography was a simple process after one had passed rather rough exams in French and German. John Kesseli (1895-1980) had taken over the examining task from Mr. Sauer. It was just one three-hour oral examination, but it was taken by invitation. I never heard of any student being so foolish as to demand to be examined, regardless of his or her time in residence. The dissertation was another matter. Sauer's monographic studies served as our model. You needed a problem, one that was likely to be redefined drastically after a season or more of field observation. Then, in the library, you pursued all angles regardless of the obstacles. In my case the obstacle was learning to read Spanish by working with eighteenth-century manuscripts. Mr. Sauer had read one to me to illustrate the nuggets not to be found in printed sources. In the end, the common lament was that one's field time had been too short and had not been used as effectively as it might have been.

Two of Sauer's comments, one published and one made to a prospective graduate student, afford an insight into Sauer's scholarly orientation. He noted that when it was published Alexander von Humboldt's *Kosmos* (Humboldt 1845-62), an effort to comprehend the physical universe, was the marvel of the age; now it is unread. On the other hand, any serious student of Mexico today and in the future must read Humboldt's less pretentious *Essay on the Kingdom of New Spain* (included in Humboldt 1807) (Sauer 1945). Put another way, the greatest discoveries of science are destined to be superseded, yet informal and insightful observations at a given place and time find a permanent place in human knowledge. To the graduate student Sauer indicated why he was happy to have become a geographer rather than a mathematician or physicist. At the frontiers of knowledge in the latter fields, each problem is a unit. A brilliant solution allows you to tackle another problem from essentially the same base. Typically, the greatest achievements are made by persons in their youth. In geography one faces a world so varied that a lifetime is too short to even see it, let alone learn much about how each place got to be the way it is. It is not unreasonable to hope that continuing observation and study will be cumulative, making possible the recognition of patterns and trends that one less experienced could not see. One can continue to become a wiser geographer.

While this paper will not attempt to analyze Sauer's intellectual development in any detail, certain continuities as well as an enlarging of interests can be noted in his published work. His early monographs were regional with a heavy geologic component, but there was more than ordinary interest in the vegetation the first white settlers saw, how it related to soil, and how settlers perceived and exploited their new environment (Sauer 1916, 1920, Sauer and Meigs 1927). His concern with the Indians was evidently not highly developed until he moved to California and began fieldwork in the American Southwest and in northwest Mexico.

At the same time Sauer was developing the Michigan Land Economic Survey (Sauer 1917, 1919, 1921, 1924). Like the Ozarks of his dissertation, the cutover lands of northern Michigan represented places where, in a generation or two, Western commercial exploitation had visibly degraded the landscape and was impoverishing communities. The fractional code for land classification was applied geography. Could we learn enough about some difficult lands to help the occupants develop a safer means of living in them? It must be added that he was designing a method; others would have to carry it out.

First in Baja California at San Fernando Velicatá (Sauer and Meigs 1927) and then more strongly in Arizona, Sonora, and northwest Mexico in general, Sauer's attention and curiosity were pulled further back into time (Sauer and Brand 1932, Sauer 1932). The Indians the Spaniards saw in the second quarter of the sixteenth century had occupied their lands a long time and had learned much about how to keep them fruitful. Collaborative learning with A. L. Kroeber (1876-1960) and his graduate students in the Berkeley Anthropology Department was readily accessible. The distinguished *Ibero-Americana* series was founded as an outlet for monographic studies dealing largely with the Indians of Middle America and their relations with Spanish invaders. Sauer contributed five studies and his students many more.

Until the middle 1930s Sauer's concern with more sustainable rural land use in the United States was conveyed through government channels (Sauer 1917, 1934a, 1934b). Thereafter, he became disaffected with a government that seemed more concerned with expanding its authority, satisfying interest groups, and in getting itself elected than in modifying contemporary economic attitudes, seeking maximum quick profits rather than obtaining sustainable yields. Beginning in 1938, a small set of elegantly written, fundamentally radical essays denounced the economic policies of all economically important countries as leading to the looting of the planet (Sauer 1938a, 1938b, 1945). Mr. Sauer was notoriously conservative politically, but these essays attracted and shook up perceptive students at both ends of the political spectrum. Their rediscovery by the radical ecologists and environmentalists of the 1960s and 1970s, well after his formal retirement, made Sauer a national figure far beyond the geographic profession (Sauer 1975, 1976).

At about the same time, in the late 1930s, Sauer began publishing on the two related themes that were central to his interest during my extended period of studentship: the origins and development of plant and animal domestication, and the tracing of human impact on the landscape throughout human time (Sauer 1936, 1944). Whether these questions were the most important issues before the world might be debated, but Sauer's enthusiasm in posing them made them seem so to his students. Seeking answers to them, in seminar reports, theses, or dissertations, proved to be effective training for developing geographers, leading them into a diversity of ancillary fields and illustrating concretely the intricate interrelationships between culture and the physical and biotic environment.

The four historical volumes dealing with early European contact with, descriptions of, and impact on the New World were accomplished well after Sauer's retirement and so had less direct effect on Berkeley geography students (Sauer 1966, 1968, 1971, 1980). They involved reexamining the main original historical sources from a geographical point of view and passing a moral (sometimes heretical, as in the case of Columbus) judgment on some actors in terms of what they did to people and landscape. They are works of serious scholarship that only an extraordinarily mature scholar could attempt. Intended primarily for historians, the books have received notice, but it is too early to estimate their impact on general historiography.

One is impressed with an intellectual career that continued to grow, adding new foci while maintaining and nourishing its earlier roots. A question that troubles many of Mr. Sauer's former students is how one person could accomplish so much: teaching both undergraduates and graduates, doing a substantial amount of detailed research, as well as writing seminal essays on a variety of topics, and chairing a department, continuously, except when he was on leave doing fieldwork, for more than thirty years. A major element was that he was able to work almost exclusively at things that interested him. His major university assignment, teaching and research, was an exploratory adventure, not a chore. Another factor was the attitude he held and somehow was able to keep holding toward university administration and politics.

For Sauer, the University of California was at Berkeley. I recall showing him a list of courses I had taken on which I distinguished those taken at U.C.L.A. from those taken at U.C.B. He asked in a vexed manner, ''What's this U.C.B.?'' He viewed it as a community of scholars, faculty and advanced graduate students, who shared mutual intellectual interests or pursued some of them independently, and who sustained themselves by training able students and recruiting the ablest of them for the community. He was happy to let U.C. President Sproul deal with the legislature and allocate the resources he obtained, making few demands on him for himself or for the department. Maintaining a substantial garage budget for fieldwork and books for the library he supported, he made continuing efforts to get government and private grants to support graduate students in the field. He made practically no effort to enlarge the faculty or acquire expensive equipment. Not fighting hard for space cost the department a bit when it was moved from Agriculture and Hilgard Halls to Giannini Hall.

The Academic Senate of the University of California is an influential if not powerful institution with an elaborate committee structure in which university academic policies are debated at length. With one exception, Mr. Sauer refused to attend meetings and so stopped being appointed to committees. He regularly chided me for excessive senate involvement whenever attending a statewide meeting gave me the opportunity for a visit. The exception was the *ad hoc* committees which consider critically each appointment at tenure and all promotions. Selection of a person for permanent membership in the Berkeley community of scholars was important enough to merit his full attention.

Perhaps Mr. Sauer's idiosyncrasy that most clearly illustrates his attitude toward administration and university politics was his attitude toward the telephone. He refused to have one in his office and would go to the department office to answer a call only if he felt like it. The secretary was instructed to learn as much from a caller as possible and then say, "I'll see if he is in." Memoranda were answered on a selective basis. When John Leighly (1895-1986) took over the chair he was shocked at the number of memoranda he was expected to respond to; obviously Mr. Sauer had chosen quite a few to ignore! Perhaps, in today's bureaucratized world not even someone with Mr. Sauer's prestige could get away with such an approach to administration without irreparable damage to his department. I do not have the nerve to try, but it is nice to think about it.

Literature Cited

Humboldt, A. de. *Voyage au régions équinoxiales du Nouveau Continent, Fait en 1799-1804, par Alexandre de Humboldt et Aimé Bonpland*. 5 vols. Paris, 1807.

Humboldt, A. von. *Kosmos: Entwurf einer physichen Weltbeschreibung*. 5 vols. Stuttgart and Tübingen: Cotta, 1845-1862.

Sauer, C. O. *Geography of the Upper Illinois Valley and History of Development*. Bulletin No. 27. Urbana: Illinois Geological Survey, 1916.

Sauer, C. O. "Proposal of an Agricultural Survey on a Geographic Basis." *Michigan Academy of Sciences, Nineteenth Annual Report* (1917): 79-86.

Sauer, C. O. "Mapping the Utilization of the Land." *Geographical Review*, vol. 8, no. 1 (1919); 47-54.

Sauer, C. O. *The Geography of the Ozark Highland of Missouri*. Bulletin No. 7. The Geographical Society of Chicago. Chicago: University of Chicago Press, 1920.

Sauer, C. O. "The Problem of Land Classification." *Annals*, Association of American Geographers, vol. 11 (1921): 3-16.

Sauer, C. O. "The Survey Method in Geography and its Objectives." *Annals*, Association of American Geographers, vol. 14, no. 1 (1924): 17-33.

Sauer, C. O. "The Morphology of Landscape." *University of California Publications in Geography*, vol. 2, no. 2 (1925): 19-54.

Sauer, C. O. "The Road to Cíbola." *Ibero-Americana*, no. 3, 1932.

Sauer, C. O. (with C. K. Leith and others). "Preliminary Recommendations of the Land-Use Committee." In *Report of the Science Advisory Board, 1933-34*, pp. 137-161. Washington, D.C., 1934a.

Sauer, C. O. "Preliminary Report to the Land-Use Committee on Land Resource and Land Use in Relation to Public Policy." In *Report of the Science Advisory Board, 1933-34*, pp. 165-260. Washington, D.C., 1934b.

Sauer, C. O. "American Agricultural Origins: A Consideration of Nature and Culture." In *Essays in Anthropology Presented to A. L. Kroeber in Celebration of his Sixtieth Birthday, June 11, 1936*, pp. 278-297. Berkeley: University of California Press, 1936.

Sauer, C. O. "Destructive Exploitation in Modern Colonial Expansion." *Comptes Rendu du Congrès International de Géographie, Amsterdam,* vol. 2, sect. 3c (1938a): 494-499.

Sauer, C. O. "Theme of Plant and Animal Destruction in Economic History." *Journal of Farm Economics,* vol. 20, no. 4 (1938b): 765-775.

Sauer, C. O. "A Geographic Sketch of Early Man in America." *Geographical Review,* vol. 34, no. 4 (1944): 529-573.

Sauer, C. O. "The Relation of Man to Nature in the Southwest." *Huntington Library Quarterly,* vol. 8 (1945): 116-125, discussion, pp. 125-130, 132-149.

Sauer, C. O. *The Early Spanish Main.* Berkeley and Los Angeles: University of California Press, 1966.

Sauer, C. O. *Northern Mists.* Berkeley: University of California Press, 1968.

Sauer, C. O. *Sixteenth Century North America: The Land and the People as Seen by the Europeans.* Berkeley: University of California Press, 1971.

Sauer, C. O. Reprint of "The Relation of Man to Nature in the Southwest" (1945) in *The New World Journal* (Berkeley), vol. 1, 1975.

Sauer, C. O. Reprint of "Themes of Plant and Animal Destruction in Economic History" (1938b) in *Co-Evolution Quarterly,* Summer, 1976.

Sauer, C. O. *Seventeenth Century North America: Spanish and French Accounts.* Berkeley: Turtle Island Foundation, 1980.

Sauer, C. O., and D. Brand. "Aztatlán: Prehistoric Mexican Frontier on the Pacific Coast." *Ibero-Americana,* no. 1, 1932.

Sauer, C. O., and P. Meigs. "Lower California Studies. I. Site and Culture at San Fernando de Velicatá." *University of California Publications in Geography,* vol. 2, no. 9 (1927): 271-302.

Sauer and "Sauerology": A Student's Perspective

Marvin W. Mikesell

I hope I can be forgiven for a cryptic and perhaps provocative title. It is, I am sure, preferable to the one that first came to my mind: "Sauer's Role in Changing the Face of the Author of this Paper"! As a former student I have read the recent spate of articles about Sauer with keen interest and an odd feeling of both elation and disappointment. That he should be portrayed as a heroic figure, larger than life, is, needless to say, not disappointing. If viewed in the context of the development of American scholarship in the twentieth century, Sauer's contribution was substantial and his reputation well deserved. Yet it is difficult for me to see him as a "saint-guru" or the author of epigrams designed to inspire a mystical conception of the quality of life on our planet. Nor was he, in my experience, the defender of a "school" of geography.

It is even more difficult to credit Sauer with explicit or novel ideas on graduate education, for the "sink or swim" policy in his department encouraged programs rather than a program. The key issue was problem formulation—students were on their own to find appropriate methods. The seminars offered by Sauer and other members of the Berkeley faculty permitted neophyte scholars to work independently on topics of mutual interest, and discussion focused on what was deserving of further study. The methodological essays published by Sauer were never reviewed in these seminars and seldom mentioned elsewhere. They were "speeches" or "editorials" and hence of little importance in a program of shared substantive inquiry. Indeed, Sauer seldom referred to *any* of his writings. What was published was finished or temporarily suspended, and present or future work was the only issue of concern. The image of Sauer laying copies of "The Morphology of Landscape" (1925), "Foreword to Historical Geography" (1941a), or "The Education of a Geographer" (1956) before a group of disciples and expounding on the doctrine of a "school" has no basis, so far as I know, in reality.

My first conversation with Sauer dealt with marine terraces, the "Pennsylvania Dutch" origin of my peculiar name, and California Indians.

144

Our last conversation (much to my distress) dealt with Barbary apes. In between we talked about a wide range of topics and may even have gossiped a bit about geography and geographers, but he never tried to define or defend a personal conception of the nature of geography. The values he cherished—knowledge for its own sake, historical perspective, cultural relativism, and the primacy of diffusion over independent invention—and his negative thoughts about academic fads, bureaucracy, social engineering, and "destructive exploitation" were both independent of and more important than the wisdom he had to offer as professor of a particular field of knowledge or chairman of what he often described as an "administrative convenience."

These remarks encourage some skeptical thoughts about the notion of a "Berkeley School." Was the enterprise founded and led by Sauer for more than thirty years a school of cultural geography, as is often supposed, or one of biogeography, historical geography, Latin-American geography, or indeed of a geography that defies definition? How can so many schools be *a* school? To be sure, some of us sought, after leaving Berkeley, to perpetuate the interests we had acquired there, and "cultural geography" was eventually used widely as a label for our efforts. Sauer himself used this expression only twice in the titles of his writings and in both cases as a label for virtually all of human geography. Whether he approved or even was well aware of the "cultural geography" proclaimed by his students is problematical. In any case, he never felt obliged to delimit or defend a subfield. The question posed initially by Harold Brookfield (1964) and later by others—"But is it *cultural* geography?"—was a problem for us, not for him.

Enough has now been said to suggest that an element of latent tension may be evident between "Sauerologists" and students of Sauer. The former, using "speeches," "editorials," and correspondence, have painted a picture that is incomplete and, in my view, not very lifelike. Cooperation between "Sauerologists" and former students is clearly needed and will have to be evident in any convincing biography. Yet both groups have serious limitations. "Sauerologists" are obliged to deal with programmatic statements and expressions of opinion that were offered with keen awareness of the effect they would have on an audience. They present as well the image of himself that Sauer was most comfortable with and sought to project. Thus to "Sauerologists" he often appears as an academic loner and champion of rural values in an America dominated by other values. Whatever may be said of Sauer he certainly was not a farmer. As the son of a college teacher (of music and languages) he enjoyed intellectual and aesthetic stimuli that were seldom evident in the rural America of his boyhood years. And it was Chicago, Ann Arbor, and Berkeley—not Warrenton or the Ozarks—that provided the setting for his scholarly career.

Sauer is also often described as an outsider unmoved by and having little influence on trends in American geography. In fact, his influence was persistent and pervasive. From his "remote position" beyond the Sierra Nevada he may have seemed and indeed often proclaimed himself to be a disinterested or disaffected member of the Association of American Geographers (A.A.G.). Yet he served as both President and Honorary President of the association and may still rank among the most quoted of American geographers. Also, the crucial first steps in the careers of his students were often fostered by the respect accorded to his letters of recommendation. If a full-length biography of Sauer is ever completed it will have to include a substantial chapter devoted to his role as a politician and entrepreneur. During most of his career serious concern and indeed shrewdness about the means necessary to achieve his ends (and those of his students) were emphatically evident. The antiestablishment sage depicted so appealingly by "Sauerologists" (and by Sauer himself) is not the person who campaigned persistently for research support, worked energetically to find jobs for his students, was twice elected President of the A.A.G., served as adviser to the Office of Naval Research and the Guggenheim and Rockefeller foundations, and was an active member of several editorial boards.

The opposition of image and reality is also revealed if one contrasts Sauer's many skeptical comments on theory and deductive reasoning—his avowed empiricist bias—with what can only be described as his flair for generalization (e.g., "The Personality of Mexico" 1941b) and even model building (e.g., *Agricultural Origins and Dispersals* 1952a). Again, in his published addresses and even more emphatically in his correspondence one can find biting comments on quantification and quantifiers. Yet he was enthusiastic about the application of statistical methods in linguistics. He was also an avid reader of highly technical literature devoted to genetics, methods of prehistoric dating, and a wide range of Pleistocene studies. In short, Sauer's "low-tech, humanistic" image is difficult to reconcile with the realities of his scientific career. To paraphrase the statement about Ratzel that Sauer offered in his "Foreword to Historical Geography": there may be more in the unknown than in the known author. For "Sauerologists," who have found rich and highly quotable ore in his "speeches" and "editorials," a bit of digging elsewhere can be recommended, especially in such works as "A Geographic Sketch of Early Man in America" (1944), "Early Relations of Man to Plants" (1947), and "Environment and Culture During the Last Deglaciation" (1948b). It is, I trust, unnecessary to remind anyone that the Sauer of "Folkways of Social Science" (1952b) is also the Sauer of "Homestead and Community on the Middle Border" (1962); and the Sauer of "The Fourth Dimension of Geography" (1974) is also the Sauer of *Colima of New Spain in the Sixteenth Century* (1948a).

These perhaps unexpected remarks do not mean that "Sauerologists" should be dismissed in favor of a panel of "expert witnesses." The picture displayed by former students and others who have had the benefit of personal acquaintance may also be incomplete or misleading. For example, my own student experience (1953-1955, 1957-1958) began at the end of Sauer's "agricultural origins" phase and was coincident with his "man's role in changing the face of the earth" phase. I think or at least hope that I could be a credible informant on these aspects of his career. But my fieldwork was in Morocco, an area he only visited and about which he had offered no provocative hypotheses. My experience, consequently, was very different from that of students who worked with Sauer in Latin America.

It is also necessary to take account of changes in Sauer's personality. In his contribution to "Geographers on Film" (see Dow 1983), J. E. Spencer remarked wistfully that the "benign old gentleman" interviewed by Preston James was not the Sauer he remembered. The same image of Sauer as a formidable and even frightening figure appears in the contribution of George Carter. Sauer for me was indeed benign. I do not recall an unkind word from him, and he impressed me as having an almost excessively cheerful view of the accomplishment of members of his "family." Moreover, by the time I arrived in Berkeley, Sauer had delegated a good deal of the reponsibility for the polishing of students to other members of the faculty. It was John Kesseli (1895-1980), acting like a Marine drill instructor, who cut new recruits down to manageable size. It was John Leighly (1895-1986), in a more gentlemanly way, who broke the news to you that you had not yet learned how to write. It was James Parsons (1915-   ) who checked your facts and figures, Clarence Glacken (1909-   ) who made you appreciate your intellectual heritage, and Erhard Rostlund (1900-1961) who encouraged thought that teaching could be a worthy enterprise. Sauer's role was inspirational and judgmental. It was in his seminars that one achieved a focus for dissertation research and a shared sense of problem. And it was Sauer again who cast the decisive vote on your results. Whether the Berkeley Geography Department functioned in this way in earlier years I cannot know. I suspect that the atmosphere was different and that Sauer himself was more active as a tutor, pace-setter in the field, and editor.

It seems paradoxical that Sauer should be acclaimed for his success in working with graduate students when his department had only an informal or implicit training program. Yet the procedure he adopted of relying almost exclusively on evening seminars was efficient from his point of view and produced results that have often been praised. After working independently on five or six topics and after convincing Kesseli that you might be more than a monolingual moron, you were invited to participate in an oral examination which, in my experience, was essentially a conversation. What you did prior to that time was your responsibility. If you had a research problem judged to be of significant interest and could discuss it intelligently with a very intelligent group of faculty members, then you were ready for your *Wanderjahr*. Little was said about research design or methodology. When you returned there would be opportunity to report on your findings, and the manuscript you produced eventually would be read with care not only by Sauer but also by several other members of the faculty. The critical standards of the department were most clearly expressed in this final phase of the apprenticeship. My manuscript was read by Sauer, Parsons, and M. M. Knight (1887-1981), an economic historian who had worked in North Africa, and then read again by Parsons plus Glacken, Leighly, and Rostlund when it was submitted for publication. The *laissez faire* attitude toward my seminar period was thus balanced by no less than seven critical readings of what I produced after returning from the field.

It is acknowledged generally that Sauer's students had good morale. Why this was true has not yet been explained adequately. That your interest was shared with Sauer was undoubtedly the most important factor. Also important, in my view, was the inspiration derived from awareness that he had confidence in you. It was not Sauer's habit to offer ''pep talks'' or worry about student anxieties. An anecdote from my experience illustrates this well. By coincidence I arrived in Spain when Carl and Lorena Sauer were touring Europe in a Mercedes they had acquired in Stuttgart and about which he was inordinately proud. We met in Madrid, drove leisurely through a good part of southern Spain and Portugal, and arrived eventually in Seville where Robert West (1913-   ) was doing archival research. The four of us then drove on to Algeçiras, parked the Mercedes, and took the ferry to Morocco.

For a week we traveled by bus in what would soon be my dissertation area. I received no advice from Sauer during the excursion on what I should do. Nor did he question me on what I was seeing or thought I was seeing. He was content to enjoy the landscape and reflect on the virtues of a non-Western civilization. Finally it was time for the Sauers and West to return to Spain and for me to remain behind in Morocco. West, always a gentleman, shook my hand and wished me well in my venture. Sauer, sensing that a comment was expected from him, hesitated for a moment and then said, ''You'll be all right; it's a good area.'' It seemed odd and disconcerting at the time that he should compliment the area rather than me. Yet that expression of fatalistic confidence was typical of his discourse with students. Although a pat on the back might have been more welcome, it was better for me to be left with the realization that I could be abandoned by my mentor in a ''good area'' and try to make sense of it according to my own agenda and in my own way. The open charge Sauer gave to me in Morocco may have been a conscious reiteration of what he had received many years earlier from his mentor. In the foreword to *Seventeenth Century North America* he recalls that experience in words I can regard as prophetic:

> After a year of graduate study at the University of Chicago I was sent in 1910 to study the Upper Illinois Valley. When I asked Professor Salisbury about the range of observations required, his reply was that this was left for me to determine and defend (1980, p. 9).

Whether I was able to make good use of my opportunity ''to determine and defend'' (Mikesell 1961) inspires melancholy thoughts that are best left unrecorded. What I can record is a sense of retrospective astonishment that a victim of "progressive education" could have survived in the Old World and perhaps old-fashioned academic environment created by Sauer in Berkeley. For most of us the first year in his department entailed a frantic effort to acquire not only knowledge of problems utterly alien to the experience of American suburbanites, but also some aspects of the wisdom acquired routinely in a *lycée* or *Gymnasium*. For me and most of the students I knew, Berkeley meant catching up as well as catching on. The best part of our experience there, marked by Sauer's barely perceptible smiles and frowns, may have been a realization that the education of a geographer can never be complete and must always entail remedial accomplishment.

As a final comment reflecting my ambivalent feeling about the relevance of "Sauerology" to the real person I had the good fortune to know, I can offer the following suggestions. First, too much attention is being paid to Sauer's "speeches" and "editorials" and not enough to his substantive work. To express the same thought more bluntly, too much attention is being paid to what he preached and not enough to what he practiced. Again, emphasis on his real or presumed humane qualities distracts attention from his accomplishment as a scholar and indeed as a scientist. I also fear that the persistent attempt to find derivative qualities in his students is inhibiting appreciation of their accomplishments. Finally and perhaps most importantly, too much attention is being directed to Sauer as a "Grand Old Man" of our discipline and not enough to the energy and vision displayed in his earlier years. The quality that best explains the remarkable influence of this remarkable man is so obvious that it is easy to overlook. That quality was enthusiasm: an enthusiasm that was persistent in his life and may, I hope, still be contagious.

Literature Cited

Brookfield, H. C. "Questions on the Human Frontiers of Geography." *Economic Geography,* 40 (October 1964): 283-303.

Dow, M. W. "Geographers on Film: The First Interview—Carl O. Sauer Interviewed by Preston E. James," *History of Geography Newsletter,* 3 (1983): 8-12.

Mikesell, M. W. "Northern Morocco: A Cultural Geography." *University of California Publications in Geographyy,* vol. 14, 1961.

Sauer, C. O. "The Morphology of Landscape." *University of California Publications in Geography,* vol. 2, no. 2. Berkeley: University of California Press, 1925.

Sauer, C. O. "Foreword to Historical Geography." *Annals,* Association of American Geographers, vol. 31 (March 1941a): 1-24.

Sauer, C. O. "The Personality of Mexico." *Geographical Revew,* vol. 31 (July 1941b): 353-364.

Sauer, C. O. "A Geographic Sketch of Early Man in America." *Geographical Review,* vol. 34 (October 1944): 529-573.

Sauer, C. O. "Early Relations of Man to Plants." *Geographical Review,* vol. 37 (January 1947): 1-25.

Sauer, C. O. "Colima of New Spain in the Sixteenth Century." *Ibero-Americana,* no. 29. Berkeley: University of California Press, 1948a.

Sauer, C. O. "Environment and Culture During the Last Deglaciation." *Proceedings,* American Philosophical Society, vol. 92 (1948b): 65-77.

Sauer, C. O. *Agricultural Origins and Dispersals.* New York: American Geographical Society, 1952a.

Sauer, C. O. "Folkways of Social Science." In *The Social Sciences at Mid-Century: Papers Delivered at the Dedication of Ford Hall, April 19-21, 1951,* pp. 100-109. Minneapolis: University of Minnesota Press, 1952b.

Sauer, C. O. "The Education of a Geographer." *Annals,* Association of American Geographers, vol. 46 (September 1956): 287-299.

Sauer, C. O. "Homestead and Community on the Middle Border." *Landscape,* vol. 12, no. 1 (Autumn 1962): 3-7.

Sauer, C. O. "The Fourth Dimension of Geography." *Annals,* Association of American Geographers, vol. 64 (March 1974): 189-195.

Sauer, C. O. *Seventeenth Century North America: Spanish and French Accounts.* Berkeley: Turtle Island Foundation, 1980.

PART FOUR

Mature Speculations

"Now This Matter of Cultural Geography"

Notes from Carl Sauer's Last Seminar
at Berkeley
edited by
James J. Parsons

Carl Sauer last offered a formal graduate seminar at Berkeley in the spring of 1964, some seven years after his official "retirement." It was listed as a seminar in "cultural geography," a favorite catch-all designation that allowed him to probe widely among the many themes that had engaged him in the course of his productive career. Joyce Endsley, who had recently given up her graduate work to become departmental secretary (although at heart still a geography student), attended those evening meetings. On at least one occasion she used her professional competence in shorthand to record, verbatim, one of these sessions when Professor Sauer held forth in his accustomed manner on the nature and promise of cultural geography as he had interpreted and practiced it. The typescript, untitled, but with the notation "Fourth meeting, March 1964" was found in the departmental files.

The casual, almost folksy style captures better than more formal writings the spirit of Sauer's thinking in these twilight years. He was then seventy-five and The Early Spanish Main (Sauer 1966) and Northern Mists (Sauer 1968) were in full gestation but had not yet seen the light of day.

It was in such small group seminars that Sauer seemed at the height of his powers. As Henry Bruman has noted, "as he spoke you could hear him think." The slow, measured speech, doubtless accompanied by the careful whittling of matches carried for an unlit pipe, would only occasionally be interrupted by an often mesmerized audience. In this case there were some thirty to forty persons in attendance of whom probably fewer than half would have been registered in the seminar and who, later in the fifteen-week period, would have given either oral or written reports on their individual research topics. Among the visitors would have been some faculty members—J. B. Jackson, the editor of Landscape magazine, then teaching at Berkeley; Dan Luten, chemist turned geographer who had recently come into the department from the Shell Development labs; and almost certainly others.

U. C. President Clark Kerr had authorized Geography to "recall" Professor Sauer at any time that it might be mutually agreeable. Since his retirement Sauer had traveled in Europe with his wife Lorena and granddaughters, and he had been a short-term guest lecturer at various times at Wisconsin, Indiana, Louisiana State, Texas A&M, U.C.- Riverside and U.C.L.A. But he was most comfortable on the friendly home turf where his working office was located on the top floor of the Earth Sciences Building, a panoramic view of the San Francisco Bay Area spread out before him. Clearly he found delight in keeping in touch with the new generation of students in the manner that this seminar afforded.

I have been trying to direct your and my attention to what are possible directions of what we call cultural geography; of things that represent the manner of life of people within an area. There are two points; they are not contrasting; they are probably somewhat supplementary. The term "cultural" I think is of importance here. And by cultural we mean something as simple and as comprehensive as a way of life. These areally expressed ways of life are subject to a good many different approaches. I think maybe this is one reason why I have never been any good as a methodologist, and I'm scared of people who make methodology the principal means by which they express their thinking. It is still one of the attractions of the geographic field that one can work in very many different ways, and I think this is a thing that is to be guarded and cultivated. I think it is fundamental to the academic life, to the intellectual life, to the University—that a person works at the things that interest him most and in which he thinks or hopes that he can develop an individual competence to express his findings and his thinking. Hence I am not pleased by any geographers who try to introduce—to impose—a particular way of working.

Now in this matter of cultural geography—and I think I should express that I am speaking somewhat autobiographically in these matters. Maybe I am making a plea—though I don't feel the necessity of making much of a plea—for the things that I have been interested in. We had, I think, a major place in introducing the term "cultural geography" to the American public, although it is an old and generally used and familiar term in all the continental countries in Europe. There is nothing original about that. Now as to these—I have spoken of three directions that I find or have found especially interesting.

The first of these is concerned with *regional geography,* to which I have always professed adherence, to which I have given some performance, though not as much as I should like to. I started in with a study that was—three studies—that are primarily regional human geography. I didn't do more of them for various reasons, one of them being that I thought I didn't know enough about how to go about regional geography to do it. Since then I have only written possibly an essay or two, one of these the title of which was suggested by an English scholar—on *The Personality of Britain* [Fox 1938]—I applied to an essay on "The Personality of Mexico," which I think was a worthwhile exercise [Sauer 1941]. Certainly nothing of methodological significance in it. But I was trying to see whether I could express in some fashion what made the life of a people in a predefined area—I didn't have to worry about boundaries—significant and characteristic, and gave to it a certain measure of originality. I think that good regional geographers have done this and some of them have been quite successful at it. I am not going to go into an example of what I think are good regional geographers. But in connection with this I did pick up from the German geographers—translated into English—a term that has since then had a good deal of significance to me and I think has been significant to the people who have worked here; and that is the term "the cultural landscape." Now it is a term that means really only a particular land—dimensions not considered; boundaries not involved—but a particular land as it appears in the occupation and the expression of the people who are in it. It is a term that has qualitative utility. And the people who have done the things that are remembered and used in regional geography are the people who had the industry and the insight to concern themselves with getting a large and a largely true expression of what such cultural landscapes are.

Now of course the French have been especially good at this, and they have been good at it for a long time. It goes back to one of the men who made French geography and started it off in this direction a long time ago, and that was [Paul] Vidal de la Blache [1845-1918]. And the French, I think, still probably write better regional geography than anybody else does. And in all of this there is—I think there are two dominant things: in the first place, an intelligible and meaningful *description,* and in the second place an *aesthetic appreciation* that they are not too reticent about expressing. I think this is the thing in which we have been rather largely lacking in this country. A Frenchman, like an Englishman, enjoys writing. He likes the savour of words—he likes the pictures that he can draw with words. And we have been pretty embarrassed about this sort of thing. I think this may be one reason why so many geographers have taken refuge in figures. It is an easy way to get yourself out of the difficulty of the conveyance of meaning by words and sentences and paragraphs. I am just throwing this out. I think one of the major motivations in these people who do a regional study is that they are trying to make others see an inhabited area as they have experienced and felt and thought about it. Okay. So this is not science in the American sense. But it is science in the older and more inclusive sense. It is an attempt to give a true picture. And that is a big claim. And this is done because of the fact that you have lived in an area. You have lived with the people. You think you can understand their outlook. And their outlook is expressed in the kind of a country that they have made and utilized. You know—I wonder what the Japanese geographers have done about this. Japan is a remarkable country in these respects. I don't read Japanese, of course, and summaries are very meager things. But the true and aesthetic appreciation of a Japanese region would be a task that might engage all the attention of a person with very great satisfaction.

Now a part of this sort of interest I think means that consciously or unconsciously the observer is getting into the question of how people live in their land: how well they live in their land—how properly they live in their land—what their ecologic relationship is to the land. And are they—well, occasionally you see a person who is old enough to use the word "harmony"—is this living "harmonious?" Inhabitant with habitat? Okay now, this is the end of the sermon.

We need more people who are willing to do this. [Mentioned important lecture tonight (Max E. Nicholson, Chairman of Nature Conservancy in Britain), and the fact that Jackson and others who usually attend seminar were not there because of it.] And I must mention Mr. J. B. Jackson. This little magazine *Landscape* that he has been putting out for, I think, a dozen years now, I think is one of the most thoughtful approaches to this question of the harmonious landscape of cultural expression that goes on anywhere. The sad thing about it is—how many members are there in the A.A.G.? 2,500?—and I just wonder what fraction of them even knows of the existence of this means of expression and exchange of communication that is available to them.

Okay, I promise now to go on to direction number two. And this is in principle again a simple theme, and it is partly cribbed from the physical geographers. And this theme is *man as a geographic (geomorphologic) factor*. It has found its way actually into some textbooks of geomorphology—at least one that I know; I think there is a Frenchman, too—but I am thinking of Herbert Louis's textbook on geomorphology[1] which has a slender chapter in it on man as a geomorphologic agent. Now this is a—at least a start. And of course we have been looking at it for a long time here. I think perhaps we [at Berkeley] have been more concerned with the geographic agency of man than any other group in this country has. And it started with a partial misapprehension. We picked up the term "natural landscape"—and this is an old theme again. It has been worked at for a long time by a good many people. The alteration of a "natural landscape" into a "cultural landscape." It took some time to find out that we didn't really know what a natural landscape was. It was a nice, neat little working idea. You begin with looking at an area before the hand of man has disturbed it. You've got a prehuman landscape, first modified, and then transformed by the agency of man.

Now where this thing of course runs into very grave difficulties is in the most obvious part of it. The first thing that man affects is the modification of the vegetation. And so you start with the natural vegetation and then you see the effects of man upon the vegetation. And then you find out that it is terribly difficult to get at the undisturbed vegetation. This is one difficulty, I think, that botanists have been in very largely. And Edgar Anderson [1897-1969] pointed it out vigorously a number of times: that the botanist likes to make his field studies—likes to do his collecting—where he is dealing with a vegetation that has not been disturbed. So he tends to go as far away a he can from towns and fields and obviously worked-over areas and collects there. Well, I had a salutary early experience here. We had in the Department of Zoology a very remarkable person by the name of Joseph Grinnell [1877-1939]. His classical studies were concerned with the native sparrows, but Grinnell is one of the founders of ecology. And after I had been here for a while I said, ''I am going to spend a month or two up in the Sierra, and I would like to get into parts of the Sierra where nature is inviolate.'' And he looked at me and said, ''I don't know of any.''

Now, this was drawing a focus that I had not had before. The pervasion and accumulation of human activity in modifying the biota had not been known to me until he began to open my eyes to it. But, and I think we still have on the part of these good field biologists a tendency to consider that the serious, the recognizable modifications of the organic world are late—are being expressed especially in the more active and more advanced societies. I gradually began to get hold of this. And this of course led me into another kind of interest and observation: I began to think about fire as a topic of interest and information. This is a situation which is still quite wide open. There are people, including myself, who suspect fire in any number of situations in the vegetated world. It is almost as true as this: that there are the simple and sturdy souls who identify vegetation with climate. And then there are the people like myself who wonder every time there is something peculiar about a vegetation whether somebody didn't set fire to it. And its simplest and most contested expression is in grasslands as against woodlands. But at any rate here is an inquiry that is relevant and it is likely to yield results in time: that man—and put it in crudest form—that whenever man began to have some technical facility—as a specific illustration, when he began to use fire—he entered the total ecology as a modifying agent. And in time perhaps became a dominant in the assemblage of living things. This is man breaking down the environmental relations upon which a completely primitive person—if we can think of that sort—in which a completely primitive person would function; this is simply in biotic terms now, considering the growing emergence of man as a suppressor, a substituter, a selector of other living things. And this of course is a tremendous subject; and it is not a subject in which I am suggesting at all—as I am not suggesting about any of these things—that the geographer is the expert. I am suggesting simply that the person who works as a geographer may have the opportunity to make contributions to such study.

Ecology, then, introduces man as an ecologic factor, as a modifier of the rest of the living environment. And there is a lot to be done in this direction, and much of it will not be done by geographers. But maybe a few things. For instance, apparently we know very, very little about microörganisms in the soil; or the extent and manner in which human activity affects these. Now it is not going to be true that the geographer will be the student who will determine the conditions under which nematodes get established and give a particular constellation to the life possibilities of a certain soil. But he may ask these questions. The economic geographer may ask them. He may ask what the effect of the applications of commercial fertilizers is on the microörganisms of the soils to which they are applied. It is an important question and it has had very little attention. Now I think this may be one of the things that this selecting man as an agent of biotic and superficial change may do. I think that very often we can ask questions that we can't answer, but we can help to get other people interested in them.

Now, the next items of course are even more obvious. Man as an agent of soil erosion. Now this is a kind of curious topic. You'd think that this would have hit any person who worked as a geographer between the eyes and hard. But it didn't. I think [John] Leighly [1895-1986] and I did the first fieldwork, certainly of any geographers, and some of the first fieldwork that was done in the United States—of what Hugh Bennett [1881-1960] and a very few people with him were doing, forgotten almost in the Soil Survey of the United States, on soil erosion and on the way in which kinds of tillage—we really didn't get beyond tillage—affected the loss of soil from the surface. We did this job in Kentucky, of the incidence of sheet wash, of gully erosion; of the causes that set it up, and the process by which it went on—this was just a little new focus in geomorphology that we were applying there. And when the larger organization was effected in the beginning of the Roosevelt Administration, as an academic spokesman I had the chance to write the recommendations for the establishment of a Soil Erosion Service (later changed to the S.C.S.). Now Hugh Bennett—a little old North Carolinian who had seen the farm of his family go down the washes in the Piedmont—had been working along these lines within the government. But he was an employee and he was not at the policy making or thinking level, supposedly. Then the Soil Conservation Service got this larger opportunity, this study of process in time, incidence of particular cause and effect, and—[this is not quotable!]—it did very well for a while. [C.] Warren Thornthwaite [1899-1963] was put in there in charge of physiographic and climatic research, and there was some nice work coming out of there. But then the engineers took over. And the S.C.S. got quite impassioned of this analytic approach. Because the engineers said: we can handle any problem; we are operators; we are not concerned about these academic questions you have developed here; give us the money and we will build the dams or we'll terrace the hillsides. So, originally a field of academic study, it was turned over to the practitioners and this kind of inquiry faded out. It has come back now in certain ways.

Now the striking thing is that so many geographers have been so blind to processes initiated by the energies of man as being of significance. There were two German geographers—one of them a very good man—who made a geomorphic study in South Africa in the Cape Colony area. It was an interesting geologic study. But they worked the whole thing out in terms of climate, intensity of rainfall, runoff, and so forth. They even went so far as to say the situation would be the same if there weren't any cattle farming, any human activity. The Germans have been rather especially blind in this respect. And I think it is partly due to the fact that they come out of an environment—out of a culture—in which these erosive activities promoted by man are minimal. The Germans on the whole have been the best geographers, but they have been among the poorest when it comes to seeing anything short of direct mechanical intervention in molding the landscape. The English haven't done much with this. The French have been the most alert again. And I don't know whether I know the full explanation for this. But a part of it at any rate is that Fenchmen have gone out for years, for decades, to work in their colonies and their colonies have been rather far-flung. The Englishmen went out as civil servants and some of them wrote some pretty intelligent articles of their experiences in the Royal Geographical Society's *Geographical Journal*. But the geographers didn't think much of the *G. J.* because it was a medium for the colonels who had come back from thirty years in India (and so forth) to give a lecture on what they saw. It wasn't very systematic.

But the French have been about the best in this direction. They went into North Africa and produced in Algeria a magnificent geographer by the name of [E. F.] Gautier [1864-1940] who wrote some of the best stuff we have about the hand of man on the land.[2] Gautier got himself interested in this question as to whether the serious destruction of the North African Uplands— whether it started with the Arabs and Mohammedans, Berbers—whether it started with the Romans, or when it started. But this was good stuff. Another interesting thing about Gautier. He looked like an Arab. He was a brown, tanned specimen. He was Professor at the University of Algiers. He was called to the Sorbonne. He declined. This was the most amazing thing, that any person in his right mind would stay in—well, technically Algiers was a colony—in a province—would stay out in the provinces when he could be in the center of life in Paris. There was another one who finally did go to Paris but got started elsewhere and who despite a sometimes rather loose rhetoric also contributed a good deal. And that was [Jean] Brunhes [1869-1930]. Pronounce it if you will. It is not a French name. But being approximately a Frenchman it had to be given a French pronounciation. Now Brunhes should be in the reading education of every geographer. And this was made very easy because for all his most important work there was an English translation made and commercially published [Brunhes 1920]. I don't know whether it is still in print or not. But now Brunhes—incidentally [he] was a student of Ratzel's and this is significant though I don't want to go into it; I don't want to tromp about too much in Clarence Glacken's territory—but Brunhes went out as a young man into French Indochina. He saw a good field there. And when Brunhes got around to formulating his *Introduction to Human Geography* he made two very important divisions. They are a bit on the rhetorical side but significant. "Destructive" occupation of the land and "constructive" occupation of the land. The chapter on destructive occupation or exploitation of the land is interesting. A significant formulation of the ways in which man reduces the possibilities—the inhabitability of the land that he is occupying. And Brunhes of course saw that these were varied—that these were not simply expressions of climate (this is the place where so many of the German geographers got stuck) but that they were expressions of different workings, different attitudes towards the land. Now Brunhes got this by going into a colonial area in which different native peoples (through commercial lures) had come to different degrees under the influence of French colonial management. But the French did have in numerous cases the wit to see that there was a grave hazard in shifting a population living in terms of its own economy—self-sufficient if you will—into production for the world market. In the same way there were French students—in this case not, so far as I know, with the exception of Gautier, professional geographers but *agronomes,* so called, which is a rather inclusive term in French—and forestry people, working in Africa, from Algiers into the Sahara, across the Sahara into the Sahel and Sudan, into the African rainforest. They did a great deal to identify these disturbances—the manner of their procedure—and they and some Belgians got around to studying the vanishing equatorial rainforest as in the process of giving way to the pressure of larger populations, more advanced techniques, and so forth; and as a consequence being not a permanent but perhaps a familiar feature that is in the process of vanishing and will be gone. And so these interests in man as a deforming agent of nature again, of course, turn right back into this question of human ecology, and in a form that now is acquiring a good deal of systematic attention. Okay, that is the second direction.

To put it in the very simplest and largely neglected form in terms of actual observation and process: on the one hand, the biotic picture; and on the other hand, the plain old physiographic picture of man as an agent of erosion, transportation, and deposition—the classical three considerations in geomorphology. But how much has been contributed in this direction?

Anybody want to come in here?

Question: What have the Russians done?

I am too old to know anything about what the Russians are doing. I do know that the Dutch have done some work on modification of vegetation in Dutch Guiana that is quite good. But I don't know. Maybe the Dutch on the whole have been somewhat too practical people, like Americans. You see, the Frenchman is culturally not too concerned about being practical. He may be practical; but it is not an important attribute.

Now the third direction—and now I'm beginning to head for home—is the question of the *form of culture.* And by forms I think I mean about what the anthropologists mean when they say "elements"; and something resembling what they mean when they say "traits." That is, I am not thinking now—I am not using the word "form" at all in the sense of the form in which a culture expresses itself. Not anything comprehensive, synoptic—but the particular items that a culture produces, uses, or, to put the thing in crudest terms, those nameable particular forms that you see in an area—the houses, the fences, fields; these items that you name, see, recognize.

Question: Do you use the word "artifact"?

Yes! I would be very willing to accept the term "the artifacts of man." And I think it is a rather good term. The archeologists have largely used this in terms of small objects.

Question: Is a clearing in a forest an "artifact"?

Yes, I think so.

Question: Are you limiting it to the concrete objects?

Ahhhhhh, that was bound to come up. No, in general. But in this particular connection I should say, yes. This, now this is a limitation of practicality just as the anthropologists have tended to distinguish between material culture and nonmaterial, sometimes spiritual or something of that sort. I say that what we need to learn something about are the grosser objects by which man has expressed himself in his habitation and activity. There is an enormous field there which, thank heavens, I don't need to get into, and this is the industrial morphology. And I have occasionally used the term "morphology" of course. The study of forms. It didn't begin—it didn't begin with geomorphology. I think it began with plant morphology. I think Goethe actually introduced the term. Goethe had curious notions about the basic plant and its parts. But the study of constituent forms—I say it is all right. This is a question of recognition of artifacts. One of the nicest books that I know is a job that [E.] Estyn Evans [1905-   ] of Belfast did on the Irish heritage in which he got into rather minute things which I am not talking about [Evans 1945]. But he got interested in the way in which people hung their doors and fastened them and the fireplace and its different structures and functions in the household. I am not going to attempt to give you a definition of what is included and what is not included in the cultural forms. I am interested in them as expressing a way of life by a particular people in a particular habitat. Now, when you start talking about forms, of course, immediately you are concerned with function—form as function. Function expresses itself out of form. Form makes possible the expression of function. Form. Function. You've got to be aware of both, and both are reflections of the same thing. Now, form, function, and the next question of course is: how did this form-function relationship come about? And you are into process. You are into origin. You are into modification with time. These are—thanks again for that word ''artifact''—these are the artifacts by which people express the adaptation of their evironment in what they know and what they want. That ought to be inclusive enough.

Now again—to take an embattled position—there are expressions of cultural experience and cultural values and so you have in them always a strong element for maintaining what you like because it is familiar, of having a thing, of doing a thing, because it is customary. It is the proper way. Tradition enters very strongly into these matters. You have a good deal of difficulty in explaining almost any house anywhere except by admitting this satisfaction. It is a proper kind of a house to live in. Now why is it a proper house? Because you think it is a proper house. You are accustomed to think it a proper house. These things may change. But while they are, they are. Contrast the attitude of the Japanese towards his house to our attitude. His house is functional. It is Japanese. He has different satisfactions that he gets out of the kind of a house that he lives in than those that we get.

Question: I imagine you are talking about single family dwellings—you are not including apartment living?

That is a good question. Apparently there are a lot of people who feel more at home in an apartment than they do in a single family house. Now, that action I think is contentment by familiarity in this type of space and dislike of shifting to a different kind. I think we build these patterns into ourselves quite readily. But I think you could perhaps also say that a form—I don't know about this—that a form begins as being functional. . . .

Here the notes trail off. Perhaps the Campanile had struck ten, or someone was to report, or term research topics were to be discussed. But the case had been made for at least three possible directions for cultural geographic inquiry—good regional description, human modification of landscapes, and the study of familiar cultural forms (culture ''elements'' or ''traits'') as visual, mappable expressions of ways of living and thinking. They had been themes that lay at the heart of the work of both Sauer and many of his students and for which he is perhaps best remembered.

Notes

1. *Allgemeine Geomorphologie,* 3rd edition, Berlin: DeGruyter, 1968. It was, of course, earlier editions to which Sauer was referring.

2. E.g., E. F. Gautier, *Le Sahara,* Paris: Payot, 1923. English edition, *The Sahara, The Great Desert,* with foreword by Douglas Johnson, New York: Columbia University Press, 1935.

Literature Cited

Brunhes, J. *Human Geography*. Translated by T. C. LeCompte, edited by Isaiah Bowman. New York: Rand McNally, 1920. A second, abridged translation by Ernest Row was published by Rand McNally in 1952.

Evans, E. E. *Irish Heritage: Its Landscape, the People, and their Work*. Dundalk: Dungalgan Press, 1945.

Fox, Sir C. *The Personality of Britain: Its Influence on Inhabitant and Invader in Prehistoric and Historic Times*. 3rd edition. Cardiff: National Museum of Wales, 1938.

Sauer, C. O. "The Personality of Mexico." *Geographical Review,* vol. 31, no. 3 (July 1941): 353-364.

Sauer, C. O. *The Early Spanish Main*. Berkeley and Los Angeles: University of California Press, 1966.

Sauer, C. O. *Northern Mists*. Berkeley: University of California Press, 1968.

Observations on Trade and Gold in the Early Spanish Main

by
Carl O. Sauer

edited by
William M. Denevan

Between 1951 and 1953, as an undergraduate at Berkeley, I took Carl Sauer's classes on Middle America and South America. These fascinating discourses on Indian ways and the Spanish Conquest were based on his intimate knowledge of the sixteenth-century sources, and more than any other event they led to my own interest in the historical geography of Latin America. Curiously, Sauer never published much of his regional material on the Caribbean, Central America, and South America until The Early Spanish Main *(Sauer 1966). This, probably his most outstanding book-length study, was his only regional treatment of the lands south of Mexico. As I look over those old lecture notes, especially for South America, I can only regret a vintage Sauer that is now lost.* [1]

In a few instances, Sauer was taped and these provide some idea of the man in the classroom. The speaking Sauer is somewhat different from the writing Sauer—informal, contemplative, less cautious and more speculative—a challenge to students to head off to library and field to seek answers to things they had not thought much about previously. One of Sauer's seminars was taped, transcribed, and edited by Robert Newcomb (Sauer 1963). "Casual Remarks" made by Sauer at a special session honoring him at the 1973 meeting of the Association of Pacific Coast Geographers were taped and later published (Sauer 1976). There is also a transcription of the 1970 film interview of Sauer done for the "Geographers on Film" series (Dow 1983).

In 1964, for part of the spring semester, Sauer was the Knapp Visiting Professor at the University of Wisconsin, Madison. At the time he was finishing the manuscript of The Early Spanish Main, *which concerns the Caribbean, adjacent mainland, and intervening sea during the period 1492-1519. He presented four public lectures related to that work and region. These presentations were taped by Professor Henry Sterling, partly for my benefit as I was in Peru at the time. The sound on two of the tapes is of excellent*

164

quality, and I have played them to my classes and seminars over the years. One of the tapes, titled "Indian Food Production in the Caribbean," was transcribed and published in The Geographical Review *(Sauer 1981). The other is presented here, as a further contribution to the spoken Sauer.*

Sauer was interested in Indian artifacts, trade, and metallurgy, and that is how he begins his lecture. This trade, in which gold was one element, was diverse and far ranging. He then spends considerable time talking about the Spaniards' interest in gold which he believed to have been the major motivating force behind the earliest exploration, conquest, and settlement of Latin America. He ends by asking where the alloys of silver and copper in the beautiful gold Indian objects of lower Central America came from, and he suggests long-distance trade as far south as Peru. Much of the material here was later treated in The Early Spanish Main, *but it is scattered throughout that book.*

Editing has often had to be heavy-handed as Sauer's sentences tended to ramble, and from time to time in his lecture he pointed to a wall map and referred to "here," or wrote a word on the board without speaking it. These places and words had to be inferred from context and in a few places omitted. The end product, however, retains the flavor of a Sauer lecture and is faithful to his intent.

In the discussion that followed the lecture, Sauer made a comment that clearly reveals his attitude towards explanation in geography. When asked about his "theory" regarding the origin and dispersal of maize, he irascibly replied: "This is no theory. I never had a theory in my life. I've had hypotheses, yes" (also see Entrikin 1984).

Sauer, I am sure, would have wished it known, as he did with the printed "Plant and Animals Exchange" seminar, that his statements here "in no way constitute 'published research' on his part" (Sauer 1963, p. i). He wanted to distinguish between informal speaking and carefully reasoned writing.

The original tape was transcribed by Judith Gunn, who also assisted with the editing. As for my part, it was good to ramble once more with Mr. Sauer.

Back in the innocent days of my youth, or the days of my innocent youth, before economics became a powerful arm of politics, trade was considered as being the result either of an exchange based upon difference of resources or on difference in skills, in art. This old designation can be applied within reason to the Caribbean area. The difference in resources is very much simplified here insofar as there is no contrast between lands that are wet and lands that are dry in this part of the world. Difference in resources is really primarily expressed in people who lived along the water and people who lived in the interior. Difference in art is, however, very emphatic.

Let me see whether I can put this into the simplest geographic terms to begin with. We have at the outset the island Arawak country [Greater Antilles], which is of the least interest of all in terms of its native trade. This is rather curious, but I suppose it may mean simply that this island area was far from the crossways of communication. We have the somewhat curious thing that connections were strong—apparently the same language was understood from end to end—and the people had rather unusually good geographic knowledge not only of the islands but lands to the south. Of course, as you probably know, the original indication of Florida was communicated to Ponce de León [from Puerto Rico]. The natives were not sticks-in-the-mud who didn't know anything about what was beyond their villages, yet there is extraordinarily little indication—the records are fairly good, verbal better than archaeological—that there was anything resembling an organized trade. The trade in fact, such as it was, was more a matter of sending presents from one place to another on one occasion or another. They had no need to exchange anything in terms of necessities. The food supply was varied and ample. Hispaniola is one area that had mineral salt instead of sea salt available to it. It was quite freely available, so that this commodity might not have been a trade item there.

One thing I think might be mentioned [about the island Arawak] is that they were among the most skilled woodcarvers, woodworkers, that we know of anywhere in the New World. The most famous illustration of this is the throne seats, the so-called *duhos*. They were carved out of heavy black wood which was identified in Spain at the time as ebony and which, very likely, may have been a native ebony. These were extraordinarily prized. If you want to be adventurous about it, this may be an extension of the idea of the throne that is reserved for the person of highest distinction. They were presented as presents. I think it is worth mentioning that when the principal Spanish entry into the west of Haiti took place, the natives really broke loose and threw the biggest kind of a show that they could think of to honor these people coming in. To the amazement of the Spaniards, they were presented, I think, with fourteen of these carved *duhos* as a great offering to show that the natives were not only peaceable but that they respected the Spaniards as the great folks from overseas. Presumably there is one of the *duhos* left in the Musée de l'Homme in Paris. I have never seen it. The Spaniards raved about them a good deal when they went back to Spain. The Indians polished out of similar material black bowls that also were considered very precious. Perhaps nowhere were the canoes ornamented and trimmed in fashions such as here. We may, I think, say that there was a highly developed art form that was appreciated, which dealt with perishable materials, and which has been lost to our present knowledge.

There was no interest, no significant interest, in metals as such [in the islands]. There was of course a little gold that came into the picture, but it was like being given a peace pipe onto which, by chance, a little piece of gold was inserted. Gold was not an object in itself. They made masks, headdresses, and girdles in which colored feathers were the most important thing and which were apparently done with great skill. They would perhaps work a little bit of gold in to tinkle or shine, but that is about the size of it. So this quite large island area, in which the mode of life was far from lacking in competence, almost disappears as a place for which there is any indication of trade. For the

Carib islands [Lesser Antilles], of course, we don't know anything, because the Spaniards got into trouble every time they tried to do something there, so those islands drop out of the picture.

When you turn to the mainland there is at once a record not only of trade but of organized trade, of trade in fairs [markets]. To use the Spanish word, which is precisely our "fair," they encountered *ferias* [along the Venezuelan coast]. This is Carib country, incidentally. The account [of Peter Martyr][2] says that the women haggled with the vendors having the stands, just as people would back home in Spain. These fairs handled goods of nearby production. One village might bring in pottery; another one might bring in some woven mats. Salt in bricks was traded into the interior for other goods which are not specified. There are few details, but enough to show that there was an exchange of goods at particular places and at stated times. This is a more environmentally differentiated region anyway. Most interestingly, this same coast, of course, had the greatest pearl beds of the New World. Pearls were traded here [Venezuela] for gold items, of which more later, that came out of the west—a coastal trade of pearls for gold.

At the northern extreme of this coast, there is the most interesting area of all in terms of trade, of which the only record is from a great canoe, or canoes, of Mayan merchants carrying goods brought from the west to be disposed of either locally or farther to the east. This is the famous incident of the fourth voyage of Columbus. The Spaniards were surprised to see quantities of cacao beans, but they didn't recognize yet that these were counters of trade— money, if you will. These Mayans were interested in copper tools, cutting tools and axes of hard copper. They were interested in cutting tools of obsidian, which were specified as to color and their size and character. There were also woven goods, woven in patterns of color, but I don't know how this color was applied. Here was a traveling store, representing pretty largely the wares that belonged to the Mayans and to parts farther west. The Mayans, of course, had no copper. As a matter of fact, neither copper nor finished obsidian can be placed east of Mexico City [Tenochtitlán], and so they must have been obtained in the area west of Mexico City in what we know at present as the Tarascan country. The amazing thing is that, with all of these items of a very strange and very interesting culture, Columbus turned aside and paid no attention to it.

There is another thing, reported from the fourth voyage, and that is the Mexican or the Meso-American trading system, the organized markets, which were probably the formal *tianguis* that were representative of all of the Meso-American high culture far to the north. Yucatán, of course, was on the periphery.

Now, there remains in between [Yucatán and Venezuela] a land where the fair disappears, the market as such is not recorded, and yet we know that there was a good deal of very active trade. I am speaking, more or less, of the country from south of Cartagena, actually from the Río Sinú through Panama over to what is now Costa Rica. This is an area which politically is interesting, because life here was organized under what the Spaniards called "little kings" or *reysuelos [caciques]*, hereditary princes with absolute authority, though the authority apparently was used with a good deal of discretion. The important point is that the prince and the house of the prince, sometimes, and quite properly, called a palace, was everything. It was the center of the life of the [Indian] state, some of which were very large. The two largest ones would have been the Sinú and in the interior to the south, Dabeiba, of which I shall speak again. Some of the states were quite small, but the prince's house was the court, if you admit that term. I mean social court and legal court. It was the center of ceremony, the place of prestige and of memories. If the account is correct, and if the man who gave the account, Pascual de Andagoya, was a good observer, what we had here was a life in which there were no taxes, no tributes. Even that long ago he thought that was kind of interesting. But the prince might call on anybody at any time to render any service. If there was a fight coming up, people might be called in. If there was wine to be made, they might be called in. If there was planting to be done, they would be called in. For this they received no pay, but at the end of any service they were given a good round of festivities, and then they went back to what they had been doing before. Now this absolutism has a particular significance, because it meant that the things that were required and stored were very largely in the possession (one can't quite say the property) of the prince of the little principality.

All of this adds up (to get back to my initial remarks) to a certain amount of quite meaningful trade in consumer goods. And here I refer again to my remark about "coast to the interior." The taking of fish and shellfish in any amount, turtles in almost any amount desired, was quite common along the coast. But in the interior there may have been shortages of protein food, and the drying, smoking, and salting was general. There are quite specific remarks about this, and this of course means the preparation on *barbacoas* [barbecues] of the foods that are to be traded to the island people. And it also means the preparation of salt, the other item that is traded in. Perhaps salt was the most important thing that was traded into the interior. Perhaps, and this has often been surmised, salt was the first important item of world trade in the history of mankind. The saltworks were at Paria where the Dutch later developed their world trade in salt. Salt was prepared off the coast of Sinú and along the west side of Panama. One gets the impression that they prepared the sea salt by putting it into shallow pans, evaporating it, raking it, and allowing it to be washed by fresh water until it was purified. These were not incidental operations; they were good-sized operations. And these are quite easy to understand, but why should there have been such extensive salt working here without a nearby backcountry that would have been a consumer of it is a somewhat different matter.

The art objects, the luxury items, what is known as "treasure," come next, and here the simple first distinction to be made is between objects that were gold and objects that were alloyed, for which the name *guanín* was picked up on the first expedition to Hispaniola and came to be the standard word. Now, the contrast between approximately pure gold and alloyed material is very important, and it confused the Spaniards for quite a while. It never was really worked out by them and constitutes the most interesting thing in the exploitation and the trade relations of the Spaniards with the natives. So I might as well turn to talking about how the Spaniards picked up this trail of golden treasure and bit by bit understood what it was, although they didn't understand it for quite a while.

May I offer a very elementary note in geology, by way of general context. The gold of the area was all secondary gold, concentrated through the weathering of bedrock and the transportation of the weathered material into a stream where, with its high specific gravity, it of course dropped out very early in the transport. It was all what we would call "placer" gold, with a little bit of it perhaps residual accumulation in the "B" horizon of deeply weathered surfaces. And it all derived from quartz veins, quartz veins carrying gold. This meant almost exclusively areas where, let us say, granitic rocks, granitic magma, because that is what we are talking about, had been intruded into an overlying cover which was then altered by heat and steam and molten rock and then penetrated by quartz veins. This was the commonest condition for finding gold concentrated. Almost exactly the same situation characterized gold occurrence in California. These quartz veins bearing gold, rotting, being washed down into the rapidly flowing streams, represented the places where one might find native gold. This sort of country occurred especially in two places on opposite sides of the central mountain range in Hispaniola: in the Trinidad Mountain country of central Cuba, and behind San Juan in Puerto Rico. When you get into the western Andes the story still remains about the same.

Now, the Spaniards didn't know anything about this. The Indians did have practical experience of this sort. Gold was something that you picked up in nuggets in streams, and so far as I know the Indians knew nothing at all about recovering fine gold. This business of washing gold, at least insofar as this part of the world was concerned, was completely unknown. If you wanted gold, you went out to a place where you knew that you might find gold after a freshet and you scrabbled for it. This, again, meant that there was not only no interest in the little-bitty gold pieces, but there was no interest in the great big ones. They didn't know what to do with the great big ones, but if you could pick up some that were convenient-sized, you had something that you could beat on and work up. Now, this is approximately the whole story of gold recovery in the islands and a good deal of the story over in the mainland. The Spaniards always called gold placers mines, but I doubt that one could say that the Indians knew anything about gold mining in terms of concentrating the native gold. What they did know was where to look for it with the best chances of getting some. And usually this was out of the rapidly flowing streams and especially after a freshet. This was the repeated formula, or in some cases in northwest Colombia they burned certain hillsides, because after they had burned the hillsides off there was a chance to see a gold-bearing lode in there and pick it out.

I think the next thing I should say is that this bespeaks a quite limited interest in gold insofar as most of these areas were concerned. This is emphatically true of the islands; it is somewhat true of the mainland. There was a limited interest in gold in Veragua [Panama]. It was rather interesting that before they went out to look for gold—and this was, again, scrabbling—they had certain ceremonials: they purified themselves, they abstained from relations with the women, and things of that sort, and then went out to get the gold they wanted to bring back and use primarily for trading purposes. But all of this is looking beyond an interest in gold as gold. This is the thing the Spaniards didn't understand, and the natives were horrified that the Spaniards were interested in this yellow stuff as such. In fact, there is a record, an approximate transcript of a scolding that a native chief in Panama gave to Balboa, who was one of the most intelligent and best of the Spaniards, as to what was wrong with the Spaniards, that they were after this yellow stuff and then when they got it they melted it down and determined how much of it was pure gold and how much of it was something else. Now, this was a senseless way of acting with regard to something that could be made into interesting and pleasing objects. So again I'm edging up on the *guanín* question.

When the Spaniards got [to the north coast of Venezuela], independent parties beginning in 1499, they at once ran into the trading of pearls for gold out of the west, and for golden objects. In fact, apparently the objects that they were most interested in were gold eagles, which of course did not mean round ones but rather three-dimensional castings of eagles. The people, Caribs, got this stuff from people [to the west] who were Arawaks. Incidentally, there is not the least indication that there was any enmity along this coast. And so the Spaniards tore off and found that these Arawaks got it from someplace farther west. The farther west they went, the more trade they found of usually cast figurines. Finally, the best of these expeditions, the Bastidas expedition, entered the Gulf of Urabá. They lost almost everything; they had an awful time getting back in their ships. But they did get back to Spain in 1502 with two chests of gold objects, which set the country aroar. Here these fellows come in from the Indies, and they've got the blamedest lot of things. Even the king got interested in it and said, "Now, I want this, and this, and this; here's a toy drum out of gold that I want." There are a number of these things mentioned: a crocodile, a turtle, apparently not much in the way of human figurines. But this Urabá business really set the Spaniards off. Actually, what had been done on these expeditions, culminating in that of Bastidas, is that they had worked in as close by sea as you could get to the centers of metallurgy—not to the centers of gold production, but to the centers of metallurgy. They didn't know yet really how great these were, but they were beginning to learn.

One of the gimmicks was that, in all of these exploitations, the king's share was a fifth, and the king had to have his share, and there was a certain share to the captain of the expedition, and so on down, shared to single shares. Now, this had the immediate, unfortunate result that when you got a lot of this stuff together the first thing that you did was to smelt it all down, find out how much gold there was in it, and then assign out the parts to the parties concerned. I don't know that a single bit of this material that was recovered in those early years is still in existence, and I might as well tell you right now that this period went on for twenty years. It may be that in Spain there is some. There happens to be quite a lot of it, but secured from elsewhere.

This business was finally cleared up by Balboa, who I think was the only person with a first-class mind who worked in any of this area during all this period. Balboa became aware that these objects were to be had everywhere, but that they came only from the east. Balboa finally decided that Dabeiba was the source of all the metallurgy. You can still find it on the map; it's not far from Antioquia in Colombia. He worked up pretty good evidence and said flatly, ''All the golden objects are made at Dabeiba.'' He then went up there [up the Río de Redes, or Río León], and he found that the metallurgy went on in this mountain locality all right, but that there wasn't any raw gold there, that it was all traded in from other parts. Undoubtedly some of it was brought in from way over at Veragua, but most of it came from the interior, from a country beyond, which was inhabited by some really tough, mean, low Indians. He didn't dare go in there, and it took more than twenty years before the Spaniards dared go. But Balboa made the situation quite clear, that the metallurgy was extremely concentrated, and that the people he saw didn't know any more about making gold eagles or anything like that than he did, but that up in the Dabeiba country smelting, alloying, casting, fusing, soldering, drawing—a metallurgic skill perhaps not surpassed anywhere else in the world—was practiced. But he couldn't do anything with it, because the major part of the supplies were traded in for fish and, according to him, for boys and girls from the coast by these Indians of fantastic repute as to their orneriness. At any rate, he couldn't go in beyond Dabeiba, and nobody else did for a good many years until the expeditions hit the Antioquia country proper, decades after. But this did establish one center of metallurgy.[3]

Another pretty intelligent fellow involved in this early period who is worth looking up is Martín Fernández de Enciso. He wrote the first thing called ''Geography in the New World'' (*Suma de geografía*), published in 1519, and a pretty fair job it was. Now, Enciso, who had a rather notable role in some of the dramatic events that took place, went into the Sinú country, and the Sinú country is the land on the east side of the Gulf of Darién. My guess would be that it is perhaps the most fantastic, ill-known Indian country in the New World. This may be where Bastidas got most of his [gold] stuff. The Sinú was another area that proved up to be a metallurgic center.

I don't want to weary you by repetition, but there were two of these metallurgic centers that were known, Dabeiba and Sinú, and *only* two, and neither of these was adjacent to a placer area. It was art, not location, that established this metallurgic business. There wasn't much use going into Sinú and staying there, because the amount of gold on hand at any one time was minimal. You had to get to the gold mines, and the Spaniards didn't do that until the mid-1530s, by which time Peru, of course, was setting 'em off. Now, what did they do? They made a very good living for a generation by getting hold of native gold objects, some of which were pure gold, the greatest part of which were alloyed, perhaps as much as three-fourths copper, some of them large, heavy objects, and how did they get them? Well, Balboa said, I'm coming to visit you and I'm giving you some nice presents, and what about this nice lot of things you have sitting on the shelf over here? That is, Balboa got it by exchange of presents of extraordinarily unequal value. Occasionally he toasted a *cacique*'s feet, or something of that sort, but on the whole he kept the colony going for a number of years with good returns by just saying let us have some of this. And the Indians were not too much upset.

Then comes the second period, when the colony becomes advertised. [The name of the colony] was a gimmick that King Ferdinand thought of—Castilla del Oro, the Golden Castille. This was an excellent Madison Avenue type of job that of course attracted people into there. This is the second period, the period that goes with Pedrarias Dávila, and this is a time of looting, of looting without mercy, one of the meanest times in New World history. It begins in 1514, and by 1520 Panama, which according to Oviedo had two million people when he got there, was a shambles. Instructions are quite legible about looting. The first place to loot is the palace of the next little *cacique* ahead, because the chances are that most of the good stuff is in the palace. The simplest thing is to do this at night. This was almost a technique. In the latter part of the night, set fire to it; then you just clean up in the morning. There wasn't much ingenuity to the way in which they did the thing. But towards the end of that period they discovered something that they hadn't noticed earlier. The *bohío* [great house of a chief] of the principality was the place in which the bodies of the ancestors, at least of the ruling family, were kept, more or less mummified, wrapped in whatever they had and with all their ornaments upon them. If one of these principalities had been going for quite a number of generations and had a good lot of ancestors up under the ceiling, all of them carrying all the gold ornaments that had belonged to them, the total amount was pretty handsome. In one specific incident in 1519, they found a *cacique,* who had evaded them for some time, dead and buried in his sepulchre. As they unwrapped him they got more than ten thousand *castellanos* of gold from this one body, and this was reported to the king. Knowing the character of the people who handled the job, if they reported ten thousand to the king you can be quite sure that there was a good deal more than that on this one body.

Now, why do I talk about this? Because for ten years Castilla del Oro kept going by looting—it was profitable, it was highly profitable. For the greater part of that time—they had already burned down or looted the *bohíos* of the *caciques*—it was profitable by hunting up graves and looting them. This is where the word *huaca* [grave] comes in first in the literature, and in later days a ghoul (a professional grave robber) in Colombia or Panama or Costa Rica was known as a *huaquero*, and it is a business that goes on to the present time.

Why can I speak with such confidence about the quality of the stuff that was obtained? Because we have museums that carry marvelous amounts of this native metallurgy. The National Bank of Colombia has handled the thing relatively simply. They have offered a fair premium over the market value of the object if it is turned in to the bank, so the National Bank has a museum of gold alloyed objects that is getting better every year. The Brooklyn Museum has a marvelous collection, largely secured from Costa Rica. This is post-Conquest as to acquisition, but pre-Conquest, this whole business, as to formation. This is, I think, a unique situation, where a colony could be established and made more and more prosperous by approximately no knowledge of gold itself. They did work a few Indians here and there a little bit, but there isn't a single gold placer field in the entire area, other than in Hispaniola, that is known at this time by name.

There is a final remark I should like to make, taking off on the wings of my imagination this time. I think there is no doubt that the great majority of this stuff was *guanín*. There was some pure gold that was smelted, but the great majority was alloyed. You can be fairly safe in saying that the *guanín* consisted of more than half of its weight in copper; perhaps three-quarters was an ordinary percentage of copper in the *guanín*. It was sometimes just a copper-gold alloy. This reduced the melting point, which I guess had utility. More commonly it had other stuff in it, and not infrequently it had silver in it. When Enciso came back from the Sinú with a modest haul from there, they melted it down. Enciso was an educated man, he was no fool, he knew what he was talking about; and he said that the curious thing about this stuff was that it was all gold and silver. There was no copper, which was the surprising thing about this particular lot, probably done on a particular commission for a wedding present or something like that. So they had gold, copper, silver, and occasionally platinum, but platinum we can write off as it only appeared in streams in the west mixed with gold. But the copper and silver are a great question. Where did they get it? They consumed a lot more copper, a lot more, than they ever did of gold, and they consumed a fair amount of silver. Now where are you going to find even small workable copper mines in Central America, much less silver mines? The volume of the stuff for the time in which it was produced amounted to a good deal. [There must have been] a distant trade, a distant trade by which the copper and the silver that was produced in Peru was carried in some fashion to [Central America]. In terms of fancy metallurgy, these people knew more than the Peruvians did or than the Mexicans did. Would some of these things trace back into Mexico? Peru I think is the better bet.

173

There is just one final note, on conch shells, the strombus shells that the Indians down in Peru and Bolivia tootled some of their mournful music on. This conch [*Strombus gigas,* the queen conch] came from [the Caribbean] and still does. Is there a connection between the giant strombus shells that have been traded since prehistoric times into the high civilization of upper Peru and a trade in raw materials [copper and silver]? Now, I am fairly obstinate on this business of the limited number of metallurgic centers, until you get farther into Colombia and Ecuador. There has been a wealth of this stuff collected, but I don't think there has ever been any indication of its fabrication to the west of Panama. So go and take a look at the [gold] collection in Bogotá,[4] or in Brooklyn, or in any of the museums that have some of these.

Notes

1. Notes on class lectures "South America" and "North America" delivered by Sauer at Berkeley in 1936, recorded by Robert Bowman, were published in 1985 by the Department of Geography, California State University, Northridge.

2. For sources referred to, as well as location maps, see *The Early Spanish Main* (Sauer 1966).

3. Sauer's thesis that sophisticated metallurgical processes were not practiced in Panama, and that the elaborate gold pieces found there had been traded in from the south, has been discussed and supported by Helms (1979, pp. 3, 146-150).

4. For a discussion of Indian gold and metallurgy in Colombia, including color plates of gold objects in the Museo del Oro in Bogotá, see Bray 1979.

Literature Cited

Bray, W. *Gold of El Dorado.* New York: American Museum of Natural History, 1979.

Dow, M. W. "Geographers on Film: The First Interview—Carl O. Sauer Interviewed by Preston E. James." *History of Geography Newsletter,* no. 3 (1983): 8-12.

Entrikin, J. N. "Carl O. Sauer, Philosopher in Spite of Himself." *Geographical Review,* vol. 74 (1984): 387-408.

Helms, M. W. *Ancient Panama: Chiefs in Search of Power.* Austin: University of Texas Press, 1979.

Sauer, C. O. *Plant and Animal Exchanges Between the Old and the New Worlds: Notes from a Seminar Presented by Carl O. Sauer,* edited by Robert M. Newcomb. Los Angeles: Los Angeles State College, 1963.

Sauer, C. O. *The Early Spanish Main.* Berkeley: University of California Press, 1966.

Sauer, C. O. "Casual Remarks," Offered at the Special Session in Honor of Carl O. Sauer: Fifty Years at Berkeley, Association of Pacific Coast Geographers, San Diego, 1973, edited by David Hornbeck. *Historical Geography Newsletter,* vol. 6 (1976): 70-76.

Sauer, C. O. "Indian Food Production in the Caribbean," edited by William M. Denevan. *Geographical Review,* vol. 71 (1981): 272-280.

PART FIVE

Student as Legacy

Domestication Process:
An Hypothesis for its Origin

Carl L. Johannessen

This postulation of the beginnings of agriculture is based on my synthesis of biological and cultural-historical information, and on my observations of, and interviews with, nonlettered people in regions relatively isolated from scientific influences. The hypothesis propounded here will enhance the focus of investigations and allow improved insight into the types and historical sequences of plants that were brought into cultivation. It will allow us to ask more relevant questions, observe more critically, make use of more data, cause others to gather their specialized information in a more systematic way, and perhaps even persuade them to publish more of their data of plant measurements instead of considering them excess detail.

I reject Harlan's (1975, p. 59) "no model" concept of how domestication began. We can work with hypotheses without losing objectivity by what Harlan and others consider to be preconceived notions.

I prefer to accept Harris's (1976, p. 311) idea: "To search for 'the facts' alone, without reference to any preconceived ideas, is neither feasible nor desirable; and the history of science suggests that major advances in understanding are more likely to be achieved when evidence is sought to test explicitly formulated hypotheses than when relatively random 'fact gathering' is undertaken."

I find that humans borrow ideas more easily than they invent new ways of acting and thinking. Therefore, the hypothesis I find most attractive is that the domestication process started when a person began to plant vegetatively the distinct and better selections from the wild and found the activity was successful; the ideas amplified and diffused from there around the world. From my hypothesis it follows that the number of species has continued to be augmented down to the present wherever the process was accepted. The diffusion process involving vegetative reproduction probably took tens of millennia before cultivation thoroughly diversified into seed crops. Diverging cultivation practices should have occurred early in the sequence, and the grand dichotomies of (*a*) vegetative cultivation leading to a broadcast seed

177

agriculture, versus (*b*) vegetative cultivation leading to horticultural treatment of seed may have occurred on separate continents and may have happened independently. There is simply no way to determine for certain; I still prefer a diffusion model and apparently C. O. Sauer did also.

As we reconstruct the beginnings of agriculture, we may assume that humans recognized very early the usefulness of plants and must have observed that plants could be reproduced vegetatively or by seed long before they tried any kind of planting. Let us hypothesize that the simplest possible way for a person to grow a known small plant in a site where it had not existed was to dig it up elsewhere and plant it in the new site. That way the incipient planter never lost sight of the object of concern. However, the odds seem low that the very first attempts at transplanting (or planting at random) led to the reproduction of a plant significantly different from the normal wild progenitor.

Humans eventually must have realized that certain plants of a species had more desirable qualities than the rest of the wild representatives of that species and that those better plants might be reproduced for their specific products. Something more impelling than merely concentrating an abundant wild resource was probably needed to induce humans to continue to plant and cultivate a grain crop, for instance. The more important reward for the perceptive cultivator who saw an unusual natural hybrid or mutant was to have the different, new plant where she could see it anytime she visited the site where the plant was growing. We can postulate that this would have stimulated the continued search for and selection of variants of the species that could have been reproduced and multiplied in the same way. I doubt the correctness of the concept that the domestication process was started by accidental planting of seed, as indicated by some authors (Harlan, DeWet, and Stemler 1976, pp. 6-8; Rindos 1980, 1984).

If seed continued to germinate in the garbage heaps of families and clans on the move, and if their moves were to significantly new ecological niches, we would be unable today to distinguish some ecologically significant changes in plants that were allowed to evolve in the new sites. The domestication question remains, "Would any of these varieties have been more useful, or would the direction of the evolution have been consistent with a true domestication change?" I think human transport in the last million years has not had enough time to induce random, useful varieties that would cause that kind of confusion of evidence to have taken place as Rindos (1984) implies. A change due to domestication is one in which only a few of the plant's features changed in a very marked way; the rest of the plant tends to remain relatively constant. This has been recognized since Darwin (1897, Vol. 2, pp. 202-205). Apple seeds remained at their wild size, while the apple pulp expanded. Maize ears have been selected consistently to grow longer, while male florets and pollen size apparently did not enlarge because no one was selecting for this. This holds true for all the plant parts recorded in the archaeological data I have analyzed.

Many scholars have thought about the domestication process and gathered their data as scientifically as possible for their era and knowledge. A few of their hypotheses are assembled here as a means of relating my current thoughts to them.

Darwin (1897, Vol. 1, p. 327) suggested that gatherers who enriched their surroundings with highly nitrogenous wastes might have sown seeds of the plants they used in these rich soils and thereby initiated the domestication process. As Hawkes (1983, p. 30) points out, Englebrecht (1916) suggested the same garbage dump hypothesis. These ideas go on echoing with Miller (1984), who gives credit to latrines as the source of organic debris. However, these scholars fail to indicate whether anything different from the half-million preceding years occurred 10,000 to 30-40,000 years ago. Hawkes (1983, p. 31) is obviously uncomfortable with this hypothesis too: "The hypothesis also does not explain why agriculture began only about 10,000 years ago . . . ." Obviously, it does not indicate that anything new happened.

V. Gordon Childe (1941, 1956) also had no real way of proving his "propinquity theory" (as referred to by Harlan 1975, p. 44) on the effects of increasing aridity, eustatic sea level rise, and the crowding of people. Childe's and Darwin's postulations have, however, stimulated thought and research.

Carl O. Sauer's (1952) hypothesis stated that sedentary fisherfolk in the semihumid tropics (who already had an abundant protein food source) vegetatively cultivated plants for dyes, glues, poisons, and tannins first, and then cultivated starch sources as a secondary and later innovation for food. This seems to me to be a much more logical way for the planting complex to have begun. The fact that some archaeologists have been unable to find these evidences suggested by Sauer has turned some of them away from his hypothesis. We must continue looking for the evidence. Also, we must not be stopped by Darwin's (1885, p. 133) other earlier suggestion that the origins of "first steps to civilization" (and agriculture) were too difficult to be solved. We should come as close as we can to a valid hypothesis on the basis of logic, if this is not directly counterindicated by data. Mangelsdorf, MacNeish, and Willey attempt to remove C. O. Sauer's input into the history of the origins of agriculture with the following:

> Sauer's hypothesis, though in many respects plausible, has little evidence to support it. True, there is some indication from the finding of griddles, presumably used in the preparation of manioc bread, that the cultivation of this plant began at a relatively early date, perhaps as early as 1000 B.C. . . . but it is doubtful that it preceded the cultivation of maize, beans and squashes in Mexico (1964, p. 429).

C. O. Sauer never said that manioc cultivation necessarily preceded other root crops such as arracacha, *Xanthosoma,* arrowroot, and potato. Nor was Sauer limiting vegetative reproduction of plants to food plants like manioc, nor the necessity of having archaeological griddles as the evidence of vegetative reproduction. Therefore, in my analysis, they have not disproved Sauer's hypothesis.

Introgressive hybridization was hypothesized by Anderson (1953) as the origin of domestication, as seeds and debris sprouted while piled on rubbish heaps of settled fisherfolk as suggested by Darwin. The introgression concept has been particularly fertile for biological studies. Anderson and Sauer were in direct, friendly communication, so it is understandable that they held somewhat similar thoughts. By the 1950s we began to get a process-oriented thought pattern emerging about this domestication research and these two thinkers certainly were prime movers. It may be that non-sedentary people other than fisherfolk began early, simplistic seed agriculture with their vegetatively reproduced crops. However, because they do not have to be shifting their camps continually, the fisherfolk could have become the best biogeographers and therefore are the most likely progenitors of the process.

When seed-reproduced plants began to be modified, the primitive seed cultivators of the New World and the Middle East were not necessarily starting their planting activities in a world devoid of experience with cultivation upon which they could base their innovative ideas. Hole (1968, p. 249), Braidwood (1967, pp. 88, 93-95), Flannery (1965, p. 1251), and Binford (1968, pp. 323-336) all seem to imply this lack of previous cultivating experience and skill in the Middle East. It seems more logical for the diffusion of such a complex and heretical technology as that of cultivation, even of vegetative parts, to have taken place slowly, and to have spread widely around the world in the tropical zone millennia prior to the effective planting of seed. This feature, I propose, is the hypothetical "precondition for seed reproduction" referred to by almost all those who have presented hypotheses on the beginnings of domestication. It is more logical than the independent development, *de novo*, of domestication all over the world only 10-12,000 years ago. Coastal fisherfolk move more easily than landlubbers today, and probably 20-40,000 years ago they did also. They surely carried this idea of planting pieces of plants widely.

The seed domestication process started when someone began to make use of seed for planting a variant of a species; for example, a grass with a nonshattering rachis. Flannery's (1965, p. 1251) hypothesis is that these early seed-planting activities were probably more successful when carried out at the margin of that species's home range, away from the concentration of the wild population. This also holds true for all the breeding types of seed crops. My further contribution to the hypothesis, as will be amplified shortly, is that the variant needed to have been from a self-pollinating species, further to isolate the genetic variant being planted from the wild population.

I am aware that Gorman (1977) thinks the rising sea level in Southeast Asia probably triggered domestication activity and agriculture, and that Solecki (1963, pp. 184-185), Byrne (1984) and others think that fluctuating climates started domestication, but they stretch credulity too far for Flannery

180

(1965, p. 1250) and for me. They seem to forget that changes in both the climate and the sea level have likely been occurring every 40-50,000 years back through the Pleistocene. Why did these changes induce domestication this last time but not previously? It is a belief in environmental determinism that I would have thought geographers especially would have discarded. MacNeish (1977, p. 796), though disavowing environmental determinism, nevertheless states that the domestication process was induced by climatic change in a complex ecological setting. Perhaps the anthropologists can accept this; I cannot.

The major ecological and biotic modifier controlled by humans was fire. Harris (1977, pp. 184-186) and others in his review postulate that perhaps fire was the stress that induced people to start domesticating. Since fire has been a relatively controlled stress factor for one-half million years, I cannot see the validity of this hypothesis. Harris (1977, pp. 186-188) goes on to review the hypothesis that competition for scarce resources by integrated communities may have triggered the development of primitive agriculturalists. He does not date this stress, but surely we humans had reached carrying capacity relatively rapidly in whatever climatic regime we occupied throughout our existence.

Harris (1977, p. 209) states that only in Australia, of the five continents with tropical climates, did the aboriginal population fail to develop agricultural systems, using instead only the cultivation of wild yams. However, Carter (1977, p. 95) has also found evidence that the Australian aborigine women put back a piece of the tuber after each harvest, and they did it with much religio-magic fervor. This is vegetative reproduction of a gathered crop! It is not necessarily domestication activity, however, until the woman initiates a pattern of returning to specific spots and reharvesting the tubers there because she remembers them to be different and of better quality, and subsequently plants them elsewhere, amplifying their distribution. This abstract insight ought to be called the start of the domesticating process. The fine line from gathering to domesticating has been a difficult insight to internalize.

Can researchers with access to traditional Bushmen informants or other gatherers please record whether these women are going back to especially good tubers that they remember at special spots? This type of investigation would be a reasonable starting activity in the discovery of the initiation of the domestication process, though it need not be limited to tubers—any plant that will grow from a piece placed in the ground will do. Just as Indian women basket weavers around Tulare Lake, California, returned to special spots on the margin of the lake for their tules—as Bill Preston (personal communication) has told me—perhaps many folk remembered where they had stuck plants in the mud. But since the Indian women were not plowing and throwing seed (Preston) it was not recognized by early explorers as a part of the domestication process.

In his literature review of the evolution of the domestication process, Anderson (1960) has summarized many peoples' thoughts on the subject in a somewhat similar way to Isaac (1970). In addition, Anderson presents the fascinating view that the major concern of ethnobiologists needs to be directed to the minor crops and especially the ornamental and flower-producing crops. He hypothesizes that in the West African source region of domestication, many people have been uninterested in ornamentals, started domesticating many currently minor crops, and in so doing provide evidence for another center of paleolithic domestication effort. The abundant use of ornamentals in Indonesia, Southeast Asia, and the New World provides a contrast to West Africa in this context. Their use could easily be a diffused trait, traded within the three homologous regions.

Harris (1977, p. 201-202) comments that in West Africa as well as in Mesoamerica the farmers have traditionally planted their crops in a horticultural manner; this may be highly significant in the stimulus-diffusion context. It should be remembered that *Dioscorea alata* was found on both sides of the Atlantic (West Africa and Trinidad) at the time of recontact by the Spaniards. How did that yam cross the Atlantic? Anyone planting vegetatively had to be planting individual reproductive units. This was probably the system used by the people who began the domestication system, and they probably did cross the Atlantic.

Innovative thoughts on the domestication process in the last decade have been supplied by Carter (1977, pp. 99, 116-129). He suggests that the spread of seed agriculture could have been a cultural stimulus-diffusion phenomenon, because he notes the pairing of similar species of plants cultivated on the various continents beginning 14,000 B.P. in Southeast Asia and spreading throughout the world by the movement of fisherfolk. He also postulates that seed agriculture reached the Middle East by 9,000 B.P., and that by 8,000 B.P. it had spread to Mexico. As he urges, we must check out these possibilities. We should observe and study and not just ignore this insight.

I propose that we have to go back much earlier than the period Carter (1977) was discussing to find the origins of planting activity. But in the modern world we can search for fisherfolk and gatherers who must have transplanted perennials and stuck pieces of branches or tubers, etc., in the ground in order for them to learn the essence of plant propagation. They could then learn which plants could be selected and maintained in their modified form. The maintenance and spread of special, human-selected, genetic qualities in specific organisms is the essence of the domestication process; without the genetic maintenance of these qualities domestication would not have occurred.

We come now to the most critical point in the history of the development of the domestication process: Why was the maintenance of selected forms begun? Who was most able to accomplish it without being killed by peers or the shaman? What rituals were performed in the beginnings of the process?

The women of traditional groups have normally collected the vegetable components of the diets. As C. O. Sauer used to postulate in his seminars, they were, therefore, most likely to have been the earliest biochemists and have known how much of which plants were needed to make dyes, drugs, medicines, tannins for fibers and hides, flavors, incenses, and poisons. With these powers of life and death, healing and hallucination, color and magical changes in strength of fibers and especially tasty food, it seems likely that we can suggest the primary movers in the process to have been women.

My colleague Ron Wixman and I have hypothesized that beliefs in a supernatural (earth goddesses') control over productiveness of the soil and the biota are central to any change from an earlier and traditional collecting mode to one of planting. The shamans (whether female or male) were in powerful positions to institute change and, as a result, it may have been rapid. Acceptance of the innovation of planting may well have been very slow by the "commoners," among gatherers, without the shamans' acceptance. As Carter (1977) suggested for Australian aborigines, the practices may have involved a rite of informing the deities of the earth that they should produce larger tubers, more intense dyes, or stronger drugs by planting examples of the product. The Indians of the American Southwest were reported to have felt that this is what they were doing with the maize seed when they planted it (Whiting 1939, p. 11).

Humans have been willing to do more things for "religious" reasons than for almost any other purpose; we can reasonably attribute the start of the domestication process to a modified form of this activity. The complexity of animal sacrifices at planting and harvest times probably served as stimulus for the spread of household pigs and chickens, and later sheep, goats, cattle, llama, and guinea pigs for "religious" use in this way. Chickens may have been saved especially for sacrifice in Southeast Asia and eaten only if special unction was required to counteract black magic powers (Johannessen 1981, 1982a, 1984). Chickens are sacrificed ritually at the time of planting by many cultures. It was simpler to kill a small tamed animal at vegetative planting time than to plow a field as a religious act, though they both may have been necessary in their respective cultures. Simplicity of explanation is reasonable in scientific theory.

Seed agriculture is composed of a set of complex traits that developed in the Middle East with the use of horned cattle to "inseminate" Mother Earth by pulling a plow (phallus) through the soil prior to broadcasting self-pollinating seed. It is not surprising that early Mediterranean people in their religions should have fixated on horned cattle in their ceremonials, since images of wild cattle were painted in central and dominant locations on paleolithic caves in southern France and Spain 20,000 years ago. The crescent-horned cattle were symbols of the crescent moon, were symbols of the 28-29 day cycles of "mamma" and moon, and of planting sequences needing oxen for plowing, but fixed by phases of the moon. Each of these symbols was integrated into cultivating in a marvelously complex way by the time broadcast seed agriculture came into its own in the Middle East. Seed agriculture with these "necessities" was so much more complex than horticultural planting of vegetative pieces that simplicity of explanation requires we accept the latter as the initiating activity.

If conditions in the New World can be used as a reasonable model, then once the domestication process was carried forward enough for seed to be used systematically, people treated the seed horticulturally as they reproduced desired plants, just as they would have had to do earlier with vegetatively reproduced plants. Each seed was planted individually, or at least was counted, as seeds were put in the same hole in the ground, since this is the way it is still done among Indians in Latin America. Others have pointed out that on an expanding frontier of seed agriculture, it was much easier to transport seed than vegetative root or stem parts for the frontier people to keep themselves in food. As stated earlier, the domestication process would have gone on most rapidly on the frontier.

In the Middle East, wheat and barley have been planted by broadcasting seed onto a prepared and stirred seedbed with the goal of mass production of the grain, not production of a horticulture crop. Perhaps this process was not as complex in the beginning, but apparently it may well have been because the cultivation practices in the use of these grains have diffused as if they started with a consistent complex. I recognize also that dibbling in wheat is at times practiced on very narrow terraces in Italy, for example.

C. O. Sauer, in seminars, used to postulate that the early seed crops may have started in a seedbed germination system. Then the seedlings (such as coconuts or rice) had to be transplanted as if the farmers were working with vegetative material. I find his postulation stimulating and probable.

I provide in this report a three-dimensional graph (Fig. 1) to assist in understanding this process. It has a time line running from the beginning of domestication on the left to the present day on the right. The vertical axis shows the increasing difficulty of maintenance of a selected feature. The other axis illustrates the increasing technology needed to reproduce the plant. The lines in the graph should be considered as exponential or logarithmic functions. Bars, representing each reproduction type, are located relative to the three axes. Lists of respective species of each reproductive type are given in Table 1.

Based on my assumptions about how long it would likely have taken for the large number of extant varieties of the vegetatively reproduced species to have mutated and to have been selected, my estimate for the lines crossing on the time axis in Figure 1 is the present, 3,000, 9,000, and 27,000 years as a likely range of ages. Lathrap (1977) has previously suggested 40,000 years ago as the likely age for the start of planting and this is not unlikely, except that he wanted to start domestication of plants with the seeds of *Lagenaria* sp., a mandatorily outcrossing, insect-pollinated plant. I find this difficult to accept, unless all that was wanted by the earliest planter was the wild product, because that is what they would have obtained. *Lagenaria* do not breed true upon seed selection.

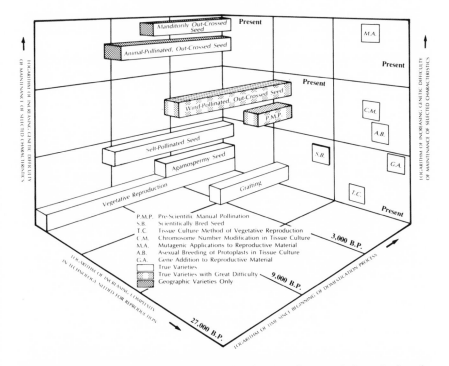

Figure 1. The systems of plant reproduction humans have used through time for domestication in relation to their genetic efficiency and technological complexity.

C. O. Sauer (1952) and Ames (1939) each proposed a much greater time depth than their contemporaries for the beginning of cultivation and domestication although they were not explicit about the time. Gorman (1977, p. 340) suggested 18,000 B. P. for the start of "incipient cultivation/tending," but much depends on the point of view on what satisfies the requirements of a significant beginning. Because I find that accrued changes in the plants are great, I think the 27,000 years is an appropriate time frame in which to hunt for evidence of beginnings.

The implication of considerable, though indeterminant, age of vegeculture in West Africa is given by Harris (1976, p. 350): "Forager populations living along forest margins and in streamside locations were the progenitors of yam domestication, and the beginnings of the process date back to Paleolithic times." For researchers who think contact between the New and Old worlds was improbable because of the inconceivability of sailing across the Pacific or Atlantic oceans, I recommend the discussions by Doran (1971) and Edwards (1965) which strongly indicate that sailing rafts and other more sophisticated craft were available to coastal people on the margins of the oceans.

185

Table 1. Typical plants in various systems of reproduction as indicated on Figure 1

Vegetative reproduction	Agamospermy seed	Self-pollinated seed	Animal-pollinated out-crossed seed	Mandatorily out-crossed seed	Wind-pollinated out-crossed seed	Grafting	Prescientific manual pollination	Scientifically bred seed and other modern innovations
Totora reed	Mango	Einkorn	Gourd	Apple	Amaranth	Apple	Date Palm	All species
Chufa	Citrus	Emmer	Squash	Pejibaye	Maize	Pear	Maize	
Water chestnut	Kentucky Blue Grass	Wheat	Avocado	Radish	Rye	Grape		
Arrowhead	Sorghum	Barley	Scarlet Runner	Petunia	Beet	Cherry		
Sweetflag		Rice	Pepper—Black	Passion Fruit	Spinach	Plum		
Barbasco		Foxtail Millet	Grape	Red Clover	Sugar Beet	Citrus		
Achiote/Bixa		Pea	Coconut	Pistachio	Buckwheat	Peach		
Tree gourd		Flax	Pennisetum	Kiwi	Swiss Chard	Avocado		
Pandanus		Lentil	Millet	Lemon	Hazelnut	Almond		
Ramón		Bean	Vicia Bean	Japanese Plum		Olive		
Black pepper		Lima Bean	Asparagus	Holly		Sour Cherry		
Hybiscus		Chili	Smyrna Fig			Macadamia		
Prickly pear		Tomato	Olive			Quince		
Canna lily		Oats	Cucumber			Pistachio		
Lotus		Lettuce	Hemp			English Walnut		
Arrowroot		Sesame	Cherry			Durian		
Arracha		Soybean	Carrot			Rambutan		
Taro		Citrus	Celery			Pecan		
Yam		Indigo	Sunflower			Nectarine		
Manioc		Apricot	Tea					
Bananas		Peanut	Black Mustard					
Plantain		Violet	Onion					
Sweet Potato		Jute	Orchid					
Potato		Eggplant	Strawberry					
Sugar Cane		Cotton	Parsnip					
Pineapple		Pigeon pea	Blueberry					
Sour orange		Jackbean	Papaya					
Grape		Kenaf	Alfalfa					
Breadfruit		Cacao						
Figs		Rhubarb						
Date Palm		Almond						
Jobo								
Garlic								
Horseradish								
Raspberry								
Gladiolus								

For those readers who have difficulty subsuming so many systems within the domestication process on Figure 1, I ask that we consider which techniques will be utilized if we start "domesticating" a plant from the wild today. I suggest we are likely to use several of the techniques illustrated on Figure 1, and that they will be appropriate to the domestication process. Therefore, my inclusion of all techniques is valid. The discussion of the individual techniques now follows.

Vegetative Reproduction

C. O. Sauer (1952) suggested that when humans began to select pieces of the maternal plant to put in the ground to develop into an entire plant, the chances were improved that they consciously could have selected for reproduction the next significant mutant that they observed. Vegetative reproduction of this type, rather than by seed, most probably occurred first because:

(*1*) The growth of a piece of rooted off-shoot, a piece of stem, rhizome, tuber, bulb, or corm is an "all or nothing" phenomenon—it reproduces "maternal" tissue of the mutant or it dies.

(*2*) The success or failure of the technique is known rapidly.

(*3*) The vegetative piece need never leave the visual field of the planter.

(*4*) It automatically maintains the desired characteristics—and this was true even when the attempted reproduction was initiated by the earliest cultivator with no previous experience.

(*5*) The technique of reproduction with automatic maintenance could remain consistent for a large number of species during this early learning process.

(*6*) Seed reproduction of the same plants rarely breeds true and 96-98 percent of the time would have given a wild progeny even when something special had been seen in the original wild parent.

(*7*) The needed advance was the discovery by the incipient cultivator of species amenable to vegetative reproduction in this "all-or-nothing" system.

After beginning vegeculture, people would have been able to learn much about cultivation practices such as methods of soil stirring, predator control, and weed control needed to allow an improved crop plant to yield effectively.

Note the sample species that are involved in vegetative reproduction (Table 1). Those crops listed under the vegetative planting column are used extensively, frequently planted widely, and are listed with the tentatively older crops first. Vegetatively reproduced plants providing containers, fibers, flavorings, drugs, medicines, and territorial or religious markers, not just edible tubers, will probably be found in the earliest archaeological strata when we learn to search for them. To find vegetatively reproduced crops may take new techniques such as the use of phytoliths (plant opal); unfortunately phytoliths do *not* of themselves demonstrate domestication. They do indicate the presence of the plant. Grass phytoliths are easily found and Turner and Miksicek (1984) have made an unsuccessful attempt at finding phytoliths in tubers. The phytoliths of maize can be differentiated from wild grasses (Piperno 1984, pp. 361-383).

More likely proof will come from finding pieces of the shell of such species as *Cresentia cujete* (tree gourd), red coloring seeds of *Bixa orellana* (achiote), fruits of *Brosimum alicastrum* (ramon), or seeds of *Piper nigrum* (black pepper). The new search needs to be made for hard plant parts (such as these species possess) that have been actively selected and in which physical characteristics have changed during the domestication process. Also, the plant parts need to be durable enough that, potentially, they can be identified in archaeological strata older than strata that contain the major, modern seed crops; the above species may provide such potential. Unfortunately, most other vegetatively reproduced plants do not have durable hard parts that were the object of selection by humans.

187

Even in the case of *Bixa,* the selection was for the dye material covering the seed, and the seed size itself may not have increased markedly in the process. I simply have not found references to archaeological *Bixa.* The great diversity in the leaf, stem, and fruit character of *Bixa* certainly indicates great age under cultivation. How far back in time does food coloring go, or how far does the use of fence plants as visual screens go? The planting of *Bixa* in fence rows in Central America is certainly widespread, and is acomplished by pushing a freshly cut branch into wet soil. The fact that only 15 years ago the apparent cultivation of the white potato (*Solanum tuberosum*) and other root crops was extended backward in time by over 4,000 years to about 10,000 B.P. bodes well for the chances of finding even older crops that were probably vegetatively reproduced (Engel 1970; Ugent, Pozorski, and Pozorski 1982, pp. 182, 191).

The speed of evolution of new forms of vegetatively reproduced crops has three very divergent regulators:

(*1*) An observable mutation must have occurred and been observed that could have been fixed as a new variety, virtually immediately, by careful collectors and planters.

(*2*) However, the rate of mutation that must have occurred and has been expressed in the vegetative plant product was a relatively slow process.

(*3*) When the plant can reproduce efficiently sexually and also is easily vegetatively planted, the speed of the process of evolution under domestication can be greatly stimulated by genetic diversity induced by sexual crossing, such as in the case of the prickly pear (*Opuntia* sp.) or tree gourd (*Crescentia cujete*).

My hypothesis is that a time depth of two or three times that of seed reproduction of crops was probably needed for cultivation of vegetatively reproduced crops to become a fixed way of life. If we do not find it in the archaeological record, we probably have looked for the wrong evidence or in the wrong ecological habitat or among the wrong people.

Perhaps we need much more study of the modern gatherers of the world to determine how greatly they modify their habitat with a few well-chosen branches that produce dye or incense or flavor, which, when stuck in the ground, will root. We need to search for a few clumps of special bromeliads, rushes, and sedges planted at the edge of the lake that provide superior fiber for basket making for years afterwards. Boundary lines made by brightly colored red leaves on *Croton* stakes stuck in the ground to root may also be ancient. The subtle difference in the archaeological records of these transplanting activities is bound to be miniscule. Nevertheless, specialists should try to discover the evidences instead of arguing that, since they cannot as yet recognize the evidence, the transplanting of vegetatively reproduced plants did not happen as the first phase in the domestication process, as logic would suggest. Let us now examine the various other systems of reproduction.

Agamospermy

Agamospermy is the substitution of vegetative for sexual reproduction within the seed that occurs in a few plants. It is a process that does not involve growth of the fused gametes as the source of the genetic constituents of the next generation (Fig. 1). Even when the gametes fuse, they may not grow; maternal tissue in the seed grows into the next generation (Stebbins 1941, 1950). This system is best known in the many varieties of *Citrus* and *Mangifera* (mangos), and it would have allowed the farmer to reproduce selected varieties and maintain them continuously when the seeds are apomictic (Table 1). It also occurs normally in Kentucky blue grass (*Poa pratensis*) (Janick 1972, p. 340), some *Malus* spp. (Sax 1959), and hawthorne (Love 1980). However, though there are several types of agamospermy, the frequency of their occurrence in the wild flora is practically insignificant and this is not a rational place to look for the origins of all domestication. We must look to the general case, normal vegetative planting, for the beginning system.

Reproduction by Self-pollinated Seed

With knowledge of vegetative cultivation securely within the technological grasp of humans, I postulate that they could have started selecting for variant forms of seed plants and experimenting with their cultivation. However, they could have been able to fix, effectively, the desired variations only in those relatively scarce wild plants that were self-pollinating. Darwin (1876) recognized this over a century ago. What he found, though apparently not registering a name for it, was the hybrid vigor of purebred lines of selfing plants when the parent plants were forced to outcross.

Current estimates indicate that only 2-4 percent of the world plant species obligatorily self-pollinate their flowers, though even these may outcross a small percentage of the time. Self-pollination, therefore, tends to produce homozygosity in only 2-4 percent of the wild plant species, and homozygosity would have had subsequent utility to the planter when an especially desirable individual plant was encountered. Even so, in the beginning of seed cultivation the planter would not know until the end of the season, or until the plant reproduced, whether she had been successful in maintaining the variant seed plant or whether she merely had the wild form.

Although his estimate may be a bit high, Brewbaker (1964, p. 30) says that almost half of the cultivated species are selfing. The implication I see in this is that humans have found it easier to incorporate into the domestication process (i.e., select, maintain, and disperse) variants of the self-pollinating wild seed plants in comparison to those that were cross-pollinated. Note that the most ancient grain crops were selfing (Table 1). For a more detailed presentation of these data refer to Table 2. It may also be that domestication and planting favored annuals and that in the selection of annualness we

Table 2. *Self-pollinated plants*

Species	Comments
Malvaceae	
Gossypium hirsutum L.	Short staple cotton—2-40% out-cross
Linaceae	
Linum usitatissimum L.	Flax, linseed—1-2% out-cross
Papilionacea	
Vicia sativa L.	Common vetch
Phaseolus vulgaris L.	Beans
P. lunatus L.	Lima beans—1-89% out-cross
Arachis hypogaea L.	Peanut—0-6% out-cross
Cajanus indicus Spreng	Pigeon pea—self-fertile, 0-65% out-cross
Cicer arietinum L.	Chick pea or gram—self-fertile, out-cross possible
Glycine max L.	Soybean
Lupinus angustifolius L.	Blue lupine
L. luteus L.	Yellow lupine—over 20% out-cross
Melilotus alba Desr.	White melilot or sweet clover—self-fertile, but insect out-cross likely
Pisum sativum L.	Field or garden pea—self-fertile, but 3% out-cross in some varieties; 24-32% in some varieties from Argentina
Indigofera sumatrana	Sumatran indigo
Lens esculenta Moench	Lentil—some out-cross
Canavalia gladiata D.C.	Sword bean—some out-cross
C. ensiformis L.	Jackbean, horsebean—some out-cross
Vigna unguiculata (L.)Walp	Cowpea—some out-cross
Compositae	
Lactuca sativa L.	Lettuce
Solanaceae	
Lycopersicon esculentum Mill	Tomato—3-12% out-cross
Capsicum frutescens L.	Bird chile—16% out-cross
C. annuum L.	Sweet chile
Nicotiana tobaccum L.	Tobacco—3-4% out-cross
Solanum tuberosum L.	Potato
S. melongena L.	Eggplant
Rosaceae	
Prunus persica L.	Peach and nectarine—largely self-fertile
P. armeniaca L.	Apricot—largely self-fertile
P. cerasus L.	Sour cherry—largely self-fertile
Tiliaceae	
Corchorus capsularis L.	Jute
C. olitorius L.	Jute
Pedaliaceae	
Sesamum indicum L.	Sesame

Source: Free 1970.

selected for selfing on the expanding frontier of agriculture, where single plants needed to have fertile seed.

Barley, wheat, foxtail millet, rice, quinoa, beans and their varieties were selected and maintained by people from their respective self-pollinating wild relatives by natural hybridization, mutation, or chromosome doubling once they were in production. Once developed, however, these grains have tended to remain relatively constant, almost automatically, because selfing increases homozygosity and decreases variability and has thus allowed a slow increase in the size of their seeds and fruits (which was slower than what occurred in the more heterozygous outcrossing species, when the latter were selected and planted consistently over many years). This difference in rate of change could only have come about after humans had learned that seed reproduction of the selfing plants was an effective way to stabilize desired traits in plants.

Tables 1 and 2 contain samples of major crops that self-pollinate their flowers. In the case of the New World beans and chili peppers, which are selfing, the archaeological records of 7,000 years ago show that they were already twice their wild size (Kaplan 1965, p. 361; Davenport 1971). This self-pollinating system results in slow change also, but maintains the differences extant. Yet, at about 7,000 B.P. the beans no longer contained parchment in the pod to cause it to open and drop seeds, and the chili fruit was already adhering to the calyx firmly and not releasing seed. How long had they already been cultivated?

We should accept the hypothesis of evolutionary specialists like Ernst Mayr (1957) in his theory of parapatric speciation that developments of new varieties are most likely on the margins of the range of the species, whether this margin be one of temperature (due to elevation or latitude) or one of moisture (due to nearness to oceans and lakes with high rainfall situations, or to deserts). These multiple ecological settings were the locations where a few individual plants with an early start, historically, might have altered significantly the genetic makeup and abundance of subsequent crop species and varieties that would have been produced. This has to be why Vavilov (1949-1950) found so many different alleles in any given crop in the most elevationally and ecologically varied terrain—not necessarily because of the species's great age under domestication at their supposed point of origin as Vavilov had hypothesized.

Since seed planting of grasses is postulated as the initiation of the domestication process by many scholars, I suggest that an additional and compoundingly difficult innovation for the preparation of the fields for grain planting and production necessarily has to be proposed as a concomitant requirement to grain domestication in the Middle East. Yet this seems to have been largely ignored. This soil preparation necessity requires a tremendous increase in work effort if we are considering the initiation of a previously undiscovered and unused system. If we remember the relatively low frequency of variety-maintaining, self-pollinating plants in the wild, the chances of finding a grass that breeds true are low. Even so, the presence of many selfing grasses, as occurs in the Sierra-Nevada foothills of California, was apparently not a stimulus to these aboriginal people to start the process. We return to the basis of the hypothesis: seeds were unlikely plant parts for starting the planting way of life.

Seed from Wind-pollinated, Outcrossing Plants

Outcrossing from wind and animal pollination has had apparent advantages for wild plants since about 96-98 percent of them utilize these two systems for fertilizing their floral parts. With wind as the main agent for dispersing pollen, the plants must necessarily grow prodigious amounts of pollen and spread it very widely during the flowering season. In agriculture, however, by use of a windbreak, or sufficient distance (40-60 meters), it is possible to isolate the source of most wind-borne pollen from maize, for example, and allow pollen from a given small population to serve as the predominant pollinator of that small population. Consistent separation of varieties was practiced with maize in forest clearings since plants could have been selected for a few consistent characteristics such as seed color and starch type. These were relatively well maintained (Johannessen, Wilson and Davenport 1970, Johannessen 1982b).

The most important wind-pollinated, outcrossing plant that has been developed and cultivated in the New World is maize. As expected, it is not as ancient as several other plant species modified in the New World. From this we can deduce five significant implications for the process as it evolved in the prescientific period:

(*1*) The absence of maize in the earliest archaeological levels substantiates my contention that under aboriginal selection and maintenance activities early improvements in the products of these plants that outcrossed were so disconcertingly slow that the early agriculturalists, effectively, could not modify them in the initial stages of the domestication process.

(*2*) The random sorting of genes from pollen blown to these outcrossing species made it very difficult for the earliest agriculturalists to improve or maintain qualities of the plant or establish varieties of these species.

(*3*) We should not expect that varieties were maintained in the earliest history of domestication from these outcrossing plants.

(*4*) Once sufficient custom, myth, and folklore arose around agricultural activities in general, we might expect that some practices were significant to the effective selection and maintenance activities for the outcrossing species, provided that selection and planting of seed were on an individual basis (horticultural) and that they continued that way.

(*5*) Once these societal controls were operative, the extreme plasticity of outcrossing maize, handled horticulturally, allowed consistent and relatively rapid changes in maize (Johannessen 1982b; Johannessen, Wilson, and Davenport 1970).

Once successful maintenance of single genes was accomplished (such as for kernel color), the efforts must have been consciously applied, through the elaboration of myth and folklore, to allow further successes with other genes and gene complexes such as for ear size in maize.

It is fair to postulate that, unless the very earliest farmers were willing to accept the benefits of close proximity to their dwellings as the *only* payoff from their efforts, they surely very soon ceased attempting to improve most seed reproducing plants with these outcrossing breeding systems. Relatively very few of them are incorporated in our prescientific seed agriculture.

Maize was not modified from the wild as early nor as rapidly as selfing chilis and beans (Kaplan 1965, MacNeish 1964, 1967, 1981). Maize was not present in the earliest archaeological sites in the Tehuacan Valley, Mexico, where evidence of agriculture has been found, nor did it become important at these locations for several millennia. The duration of this early domestication process is to be measured in millennia. It was a constant, slow change, not a revolution, and it obviously involved a tremendous amount of hard work and experimentation.

Reproduction by Outcrossing, Animal-pollinated Seed

Plant species pollinated by animals such as insects (Tables 1 and 3) cannot be isolated effectively from the pollen of another plant of the same species in a small region with only windbreaks or short distances. In comparison to the effectiveness of wind-blown pollen over kilometer-long distances (Table 1), birds, bats, bees, and insects in general are able to travel significantly greater distances, and they are more effective in cross-pollinating plants such as squashes. For perhaps millennia, gourds that were planted and grown were essentially wild plants because they were insect-pollinated. One seed (or a hundred) on the garbage dump would have produced many usable "wild" gourd shells, as Anderson (1952) pointed out. The process of regional differentiation for any given squash species develops as a result of genetic drift only with intervening distances of many kilometers because the bees and wasps keep the pollen well mixed from one plant to its wild neighbors. The manditorily outcrossing *Bactris gasipaes* (pejibaye palm) is insect pollinated by flies, bees, and wasps, and since most of these trees are planted from seed, most of the fruit is poor quality. Pejibaye can also be planted vegetatively from shoots cut from the base of the tree. (In the last decade it has been successfully reproduced in tissue culture also.) These vegetatively reproduced forms will augment the quality of this crop rapidly as it no doubt aided the process somewhat in the past. Without modern tools it was difficult to cut shoots off centuries ago in such a way that they could grow (Johannessen 1966).

Table 3. *Insect-pollinated plants*

Species	Comments
Cruciferae	
Raphanus sativus L.	Radish—self-sterile
Cochlearia armoracia L.	Horseradish
Brassica alba L.	White mustard—low selfing
Hibiscus cannabinus L.	Kenaf
Sterculiaceae	
Theobroma cacao L.	Cacao
Rutaceae	
Citrus spp.	Citrus—highly variable, many will self, agamospermy typical
Vitaceae	
Vitis spp.	Grape—highly variable, many will self, increases fruit set
Anacardiaceae	
Mangifera indica L.	Mango—out-cross but agamospermy typical
Anacardium occidentale L.	
Papilionaceae: *Medicago/Trifolium*	
Medicago sativa L.	Alfalfa—self-fertile
Trifolium pratense L.	Red clover—self-sterile
Trifolium repens L.	White clover—self-sterile
T.r. Latum L.	Ladino clover—self-sterile
T. alexandrinum L.	Egyptian clover—self-sterile
T. hybridum L.	Alsike clover
T. fragiferum L.	Strawberry clover—self-sterile, but out-cross
Vicia faba L.	Broad bean—self-fertile, but out-cross
V. villosa Rothe	Hairy vetch—self-sterile
Phaseolus multiflorus Willd	Scarlet runner bean—self-fertile, but out-cross
Coronilla varia L.	Crown vetch—self-sterile
Indigofera arrecta Hochst	Java indigo
Lotus corniculatus L.	Bird's foot trefoil
Dipteryx odorata Willd	Tonka bean
Lablab niger Medik	Hyacinth bean
Grossulariaceae	
Ribes grossularia L.	Gooseberry—self-fertile and out-cross
R. rubrum L.	Red currant—self-fertile and out-cross
R. nigrum L.	Black currant—self-fertile and out-cross
Myrtaceae	
Feijoa sellowiana Berg.	Feijoa—mainly out-cross
Psidium guajava L.	Guava—self-fertile, 35% out-cross
Passifloraceae	
Passiflora edulis Sims	Passion fruit—self-fertile

Table 3. Insect-pollinated plants (Cont.)

Species	Comments
Curcurbitaceae	
Curcurbita maxima Duch	
C. mixta Pang	
C. moschata Duch	
C. pepo L.	
Cucumis melo L.	Muskmelon, canteloupe and sweet melons—self-fertile and interfertile
Colocynthis citrullus L.	Watermelon
Umbelliferae	
Daucus carota L.	Carrot
Pastinaca sativa L.	Parsnip
Foeniculum vulgare Mill.	Fennel or saunf
Rubiaceae	
Cinchona calisaya Wedd.	
C. ledgeriana Moens	
C. officinalis L.	Quinine—out-cross
C. succirbur Pav.	
Compositae	
Helianthus annuus L.	
Vacciniaceae	
Vaccinium spp.	
V. *myrtilloides* (2X)	Blueberries
V. *angustifloium* (4X)	
V. macrocarpum L.	Cranberry—selfing possible
Chenopodiaceae	
Beta vulgaris L.	Beet—late opening of stigmas
Spinacia oleraceae L.	Spinach—dioecious plants
Polygonaceae	
Fagopyrum esculentum Moench	Buckwheat
Rheum rhaponticum L.	Rhubarb
Lauraceae	
Persea americana Mill	Avocado—possible self-pollinate as accident
Cinnamomum zeylanicum Breyn	Cinnamon
Moraceae	
Ficus carica L.	Fig, San Pedro, Smyrna, caprifig (wild)
Rosaceae	
Prunus domestica L.	Plum
Pyrus malus L.	Apple
Pyrus communis L.	Pear
Prunus avium L.	Sweet cherry

Table 3 continues on page 196.

Table 3. Insect-pollinated plants (Cont.)

Species	Comments
Prunus amygdalus Bartsch L.	Almond
Fragaria x ananassa Duchesne	Strawberry—also self-fertile
Rubus idaeus L.	Raspberry ⎤
R. strigosus Michx	Raspberry ⎮ Most out-cross
R. occidentalis L.	Raspberry ⎰ Self-pollinating forms selected
R. fruticosus L.	Blackberry ⎦
R. caesius L.	Dewberry
Lilaceae	
Allium cepa L.	Onion
Asparagus officinalis L.	
Papaveraceae	
Papaver somniferum L.	Oil poppy, opium poppy—self- and out-cross
Theaceae	
Camellia sinensis (L) O. Kuntze	Tea
Bombacaceae	
Ceiba pentandra Gaertn	Kapok—self- and out-cross
Sapindaceae	
Litchi chinensis Sonn	Lychee—slight selfing
Caricaceae	
Carica papaya L.	Pawpaw
Convolvulaceae	
Ipomoea batatas L.	Sweet Potato—reproduces vegetatively
Piperaceae	
Piper nigrum L.	Pepper
Myristicaceae	
Myristica fragans Houtt	Nutmeg, mace—insect- and wind-pollinated
Orchidaceae	
Vanilla planifolia And.	Vanilla
Bromeliaceae	
Ananas sativus Schult f.	Pineapple—self-sterile
Musaceae	
Musa sp.	Banana ding—parthenocarpic, reproduced vegetatively
Palmae	
Phoenix dactylifera L.	Date—reproduced vegetatively

Source: Free 1970.

The additional activity that the nonscientific farmers had available, which surely assisted the improvement process, was that of roguing or killing the undesirable individual plants. Pejibaye trees that produced low quality fruit were simply harvested in their entirety and the heart of the palm eaten; this terminates its production of low-quality pollen. Squash plants are sampled when the first young fruits develop, and if they are bitter, the plant is destroyed. This, too, eliminates poor pollen.

Outcrossing did not lead to major crops early in the domestication process. The fact that breadfruit, sugar cane, grapes, dates, bananas, figs, berries, etc., are outcrossings is beside the point, since all are primarily reproduced by vegetative cuttings, not seed! The age of the other self-sterile, mandatorily outcrossing, animal-pollinated crops (Table 1) is probably not great and you may notice that many are perennial tree crops reproduced by grafting—rarely by seed in modern horticulture (Tables 1 and 3).

From the meager evidence available to us it appears that most weedy species are outcrossing plants with great genetic plasticity. It is because of this labile characteristic that weedy plants have been the constant companions of humans since we began to make trails and to set and control fire. When humans started controlling vegetation by planting in clearings, those clearings provided the weedy species with ecological niches in tremendous quantity. This fact may suggest that for the weedy plants with useful characters either: (*1*) those features were actively selected continually over millennia, just as if the species could have been changed as though it were a selfing crop (yet in a single generation the outcrossers would not have rewarded or reinforced the planter's efforts, since the selection efforts alone cannot maintain the desired feature on a one generation basis); (*2*) the species may simply have remained as a sun-loving, highly adaptable weed; or (*3*) the species may have been such a good mimic, such as rye was in barley and wheat fields, that it self-domesticated without the farmer's selection by supplying ripe seed at the time of the wheat and barley harvest which tended to be done *en masse*.

Grafting

Grafting had developed in the Middle East and was used in Greco-Roman times, though its exact beginning is not well known (Fig. 1). The process allowed farmers to obviate the problems of outcrossing dilution of the higher quality plants present as mutants or chance hybrids by maintaining them in the grafts. But grafting is technologically more difficult to accomplish than the mere rooting of a stem or tuber in simple vegetative reproduction or by use of seed; therefore it is moved to the right on the diagram. Yet it is the system used in almost all our modern orchards; it simply is not very ancient.

Prescientific Manual Pollination

At least two examples of manual pollination of dates and maize are known prior to the knowledge of genetics. Early in Egyptian horticulture, male date flowers were hung in the date racemes to outpollinate and provide special qualities to the date flesh. Maize in the early nineteenth century was modified by hand pollination of corn silks, and new varieties were made without the breeders knowing how the pollination process worked. The location on Figure 1 shifts accordingly.

Scientific Breeding

As indicated previously, great changes have been wrought once we learned about the significance of the control of the pollination processes. Any species can now be studied and bred; none is now really intractable. We can place any plant (or animal) in a domestication process if we are willing to finance the work. We should not be limited by what early farmers found to be unrewarding work.

Tissue Culture

Because azenic (germ free) tissue culture is essentially vegetative reproduction, it allows exact replication most of the time (Fig. 1). Some possiblility of mutation occurs in this recent, higher technological system; therefore, tissue culture is shifted up on the vertical or "maintenance" axis and way over on the "complexity" axis. With research any species can be reproduced, and hundreds have already been replicated with this method.

Chromosome Modification and Mutation Breeding

Chromosome modification requires increasing technology from simple azenic tissue culture, but means of doubling or halving chromosome numbers are available now with colchicine and other chemicals (Jackson 1976, p. 219). By definition, treatments such as hybridizing after x-ray irradiation, called mutation breeding, change the type of plant by mutagenic applications to reproductive material (Fig. 1). Therefore, "maintenance" takes on new and different significance. It does not maintain the old features, but the higher technology does induce change and allow possible selection for rapid improvement of the product. Therefore, the location on Figure 1 is placed higher and to the right. It should also be noted that tetraploids are more easily maintained in self-pollinating crops (Jackson 1976, p. 219); they make "better" plants.

Genetic Engineering

Both protoplast fusion or asexual breeding and gene addition to proto-plasts can be subsumed under the concept of genetic engineering to allow us to continue this domestication process with increasing technology. Potentially we can continually improve crops and bring new species into the process. Problems are attendant to all these breakthroughs. Wisdom is needed. Humans and all biota are now on the edge of a vista as great as that created by the start of the domestication process itself and equally afflicted with unknowns. All wild organisms are potential genetic resources; we dare not lose them by extermination or fail to diversify our crops by adding heretofore uncultivated crop species.

Concept of Varieties

True variety status is only possible (under prescientific technology) for the vegetative, agamospermous, and self-pollinating species. Since almost all *in vitro* tissue culture reproduction is vegetative, true varieties are provided with this new technology once selections are made and the numbers multiplied in tissue culture. Modern genetic control can also produce true varieties after many generations of controlled breeding of outcrossing species. Except for the case of maize with its tremendous load of cultural traits designed for attempted maintenance of its varieties (in spite of its normal tendency to be cross-pollinated by wind), only geographic varieties were possible with the outcrossing plants prior to genetic knowledge a century ago.

Proposals for Further Research

The hypothesis of the origin of domestication proposed here has certain aspects that should allow its application in the process of choosing a new species to be brought into domestication. The outcrossing species of the world are the biotic potential for the modern world's development because, prior to the last hundred years, it has remained difficult for the primitive farmers and for the early "scientists" to maintain selected varieties. We can improve these outcrossing species now as a result of our knowledge of genetics and breeding control.

Until recently, we tended to believe our historical antecedents had tapped all the "good" plants already. It is my contention that our ancestors started with plant forms not greatly different from the appearance of the wild plants we see now. They simply had most of their success with the selfing wild forms. This leaves 96-98 percent of the flora (and fauna) from which we can choose future plants (or animals) to enter into the domestication process.

Consequences for Policy

The consequences of this study are that we now have a framework within which: (*1*) to place the process of domestication in the continuum from the historical to the present time; (*2*) to cause questions to be asked during archaeological investigations that otherwise might be overlooked; (*3*) to stimulate a search for plant parts (such as plant opal) that could indicate early vegetative reproduction of plants in older archaeological sites; (*4*) to understand why so "few" plants of the total number in the world's flora have been brought into the domestication process so far; (*5*) to give us the confidence to consider any species, including the remaining outcrossing plants, as a potential crop plant; (*6*) to demonstrate this potential value of wild biota and cause us to preserve the world's gene pool from destruction through forest clearing or overgrazing currently devastating 300 million years of evolution; and (*7*) to understand the causes for the differences in the rates of change of the parts of the plants under selection pressure when these plant remains are found in archaeological sites because homozygous (self-pollinating) plants are so restricted in their genetic plasticity. They have been changing more slowly than the heterozygous plants once the latter were started in the domestication process. Therefore, we should scrutinize more of the 96-98 percent of the world's outcrossing plants and bring many of them into the domestication process with the assurance that their plasticity will allow improvements in them much more rapidly than occurred during the last 9-10 millennia in the case of the majority of the self-pollinating seed crops, or the 20-30 millennia for vegetatively reproduced crops.

Acknowledgments

I am grateful to Doris Johannessen, Maxine Thompson, Jonathan Sauer, Ed Price, Christopher Salter, Ruth LeFevre, Martha Burns, Henry Lawrence, Fred Wilson, Lance Sentman, David Brenner, Anne Parker, Howard Horowitz, several students, and the clerical staff of the Department of Geography, University of Oregon, for their helpful suggestions on the improvement of this manuscript, . It was first given in honor of Professor Carl O. Sauer who stimulated my interest in the subject.

Note

A shorter, earlier version of this paper was given at a session in honor of C. O. Sauer at the A.P.C.G. meeting in San Diego, California, June 14-15, 1973.

Literature Cited

Ames, O. *Economic Annals and Human Cultures*. Cambridge: Massachusetts Botanical Museum of Harvard University, 1939.

Anderson, E. *Plants, Man and Life*. Berkeley: University of California Press, 1952.

Anderson, E. "Introgressive Hybridization." *Biological Reviews of Cambridge: Philosophical Society*, vol. 28 (1953): 280-307.

Anderson, E. "The Evolution of Domestication." In *The Evolution of Man: Man, Culture and Society*, edited by S. Tax, pp. 67-84. Chicago: University of Chicago Press (Vol. 2 of *Evolution after Darwin*), 1960.

Binford, L. R. "Post-Pleistocene Adaptations." In *New Perspectives in Archaeology*, edited by S. R. Binford and L. R. Binford, pp. 323-336. Chicago: Aldine, 1968.

Braidwood, R. J. *Prehistoric Men*. 7th edition. New York: William Morrow Co., 1967.

Brewbaker, J. L. *Agricultural Genetics*. Englewood Cliffs, New Jersey: Prentice-Hall, 1964.

Byrne, R. "Climatic Change and the Origins of Agriculture." *A.A.G. Program Abstracts*, R. J. Earickson (compiler), p. 236. Washington, D.C.: Association of American Geographers, 1984.

Carter, G. F. "A Hypothesis Suggesting a Single Origin of Agriculture." In *Origins of Agriculture*, edited by C. A. Reed, pp. 89-135. The Hague: Mouton Publishers, 1977.

Childe, V. G. *Man Makes Himself*. Revised edition. London: Thinker's Library, 1941.

Childe, V. G. "The New Stone Age." In *Man, Culture and Society*, edited by H. L. Shapiro, pp. 94-110. New York: Oxford University Press, 1956.

Darwin, C. R. *The Effects of Cross and Self Fertilization in the Vegetable Kingdom*. London: John Murray, 1876.

Darwin, C. R. *The Descent of Man and Selection in Relation to Sex*. 2nd edition. London: John Murray, 1885.

Darwin, C. R. *The Variation of Animals and Plants Under Domestication*. 2 vols. New York: D. Appleton Co., 1897.

Davenport, W. A. "Domestication of the *Capsicum* Peppers." Unpublished Ph.D. Dissertation. Eugene: Geography Department, University of Oregon, 1971.

Doran, E. D. "The Sailing Raft as a Great Tradition." In *Man Across the Sea*, edited by C. L. Riley, J. C. Kelly, C. W. Pennington, and R. L. Rands, pp. 115-128. Austin: University of Texas Press, 1971.

Edwards, C. R. "Aboriginal Watercraft on the Pacific Coast of South America." *Ibero-Americana*, vol. 47 (1965): 1-138.

Engel, F. "Exploration of the Chilca Canyon, Peru." *Current Anthropology*, vol. 11 (1970): 55-58.

Englebrecht, T. H. "Uber die Entstehung einiger feldmassig angebauter Kulturpflanzen." *Geographische Zeitschrifte*, vol. 22 (1916): 328-335.

Flannery, K. V. "The Ecology of Early Food Production in Mesopotamia." *Science*, vol. 147 (1965): 1247-1256.

Free, J. B. *Insect Pollination of Crops*. New York: Academic Press, 1970.

Gorman, C. "*A Priori* Models and Thai Prehistory: A Reconsideration of the Beginnings of Agriculture in Southeastern Asia." In *Origins of Agriculture*, edited by C. A. Reed, pp. 321-355. The Hague: Mouton Publishers, 1977.

Harlan, J. R. *Crops and Man*. Madison: American Society of Agronomy, 1975.

Harlan, J. R., J. M. J. De Wet and A. B. L. Stemler. "Plant Domestication and Indigenous African Agriculture." In *Origins of African Plant Domestication*, edited by J. R. Harlan, J. M. J. De Wet, and A. B. L. Stemler, pp. 3-22. The Hague: Mouton Publishers, 1976.

Harris, D. R. "Traditional Systems of Plant Food Production and the Origin of Agriculture in West Africa." In *Origins of African Plant Domestication*, edited by J. R. Harlan, J. M. J. De Wet, and A. B. L. Stemler, pp. 311-356. The Hague: Mouton Publishers, 1976.

Harris, D. R. "Alternate Pathways Toward Agriculture." In *Origins of Agriculture*, edited by C. A. Reed, pp. 173-249. The Hague: Mouton Publishers, 1977.

Hawkes, J. G. *The Diversity of Crop Plants*. Cambridge: Harvard University Press, 1983.

Hole, F. "Evidence of Social Organization from Western Iran, 8,000-4,000 B.C." In *New Perspectives in Archaeology*, edited by S. R. Binford and L. R. Binford, pp. 245-265. Chicago: Aldine, 1968.

Isaac, E. *Geography of Domestication*. Englewood Cliffs, New Jersey: Prentice-Hall, 1970.

Jackson, R. C. "Evolution and Systematic Significance of Polyploidy." *Annual Review of Ecology and Systematics*, vol. 7 (1976): 209-234.

Janick, J. *Horticultural Science*. 2nd edition. San Francisco: W. H. Freeman and Co., 1972.

Johannessen, C. L. "The Domestication Process in Trees Reproduced by Seed: The Pejibaye Palm in Costa Rica." *Geographical Review*, vol. 56 (1966): 363-376.

Johannessen, C. L. "Folk Medicine Uses of Melanotic Asiatic Chickens as Evidence of Early Diffusion to the New World." *Social Science and Medicine*, vol. 15 (1981): 427-434.

Johannessen, C. L. "Melanotic Chicken Use and Chinese Traits in Guatemala." *Revista de Historia de America* (1982a): 73-89.

Johannessen, C. L. "Domestication Process of Maize Continues in Guatemala." *Economic Botany*, vol. 35 (1982b): 89-99.

Johannessen, C. L. "Distribution and Medicinal Use of the Black-Boned and Black-Meated Chicken in Mexico, Guatemala and South America." *National Geographic Society Research Reports for 1976,* vol. 17 (1984): 493-495.

Johannessen, C. L., M. R. Wilson, and W. A. Davenport. "The Domestication Process of Maize: Process or Event?" *Geographical Review,* vol. 60 (1970): 393-413.

Kaplan, L. "Archaeology and Domestication of American *Phaseolus* (Beans)." *Economic Botany,* vol. 19 (1965): 358-368.

Lathrap, D. "Our Father the Cayman, Our Mother the Gourd: Spinden Revisited, or a Unitary Model for the Emergence of Agriculture in the New World." In *Origins of Agriculture,* edited by C. A. Reed, pp. 713-751. The Hague: Mouton Publishers, 1977.

Love, R. M. "Insect Feeding on Indigenous and Introduced Hawthorns." Unpublished Ph.D. Dissertation. Eugene: Biology Department, University of Oregon, 1980.

MacNeish, R. S. "The Origins of New World Civilization." *Scientific American,* vol. 211 (1964): 29-37.

MacNeish, R. S. "Mesoamerican Archaeology." In *Biennial Review of Anthropology,* edited by B. J. Siegal and A. B. Beals, pp. 306-331. Stanford: Stanford University Press, 1967.

MacNeish, R. S. "The Beginning of Agriculture in Central Peru." In *Origins of Agriculture,* edited by C. A. Reed, pp. 753-801. The Hague: Mouton Publishers, 1977.

MacNeish, R. S. "Tehuacan's Accomplishment." In *Supplement to the Handbook of Middle American Indians,* vol. 1, *Archaeology,* edited by J. Sabloff, pp. 31-47. Austin: University of Texas Press, 1981.

Mangelsdorf, P. C., R. S. MacNeish, and G. R. Willey. "Origins of Agriculture in Middle America." In *Handbook of Middle American Indians,* vol. 1, *Natural Environment and Early Culture,* edited by R. West, pp. 427-445. Austin: University of Texas Press, 1964.

Mayr, E. "Species, Concepts and Definitions." In *The Species Problem,* edited by E. Mayr, pp. 1-22. American Association for the Advancement of Science, Publ. 50. Washington, D.C., 1957.

Miller, D. L. "The Latrine Area Hypothesis: Endozoochorous Dispersal, Unconscious Selection and the Possible Origin of Seed Agriculture." *A.A.G. Program Abstracts,* R. J. Earickson (compiler), p. 235. Washington, D.C.: Association of American Geographers, 1984.

Piperno, D. R. "A Comparison and Differentiation of Phytoliths from Maize and Wild Grasses: Use of Morphological Criteria." *American Antiquity,* vol. 49 (1984): 361-383.

Rindos, D. "Symbiosis, Instability, and the Origins and Spread of Agriculture: A New Model." *Current Anthropology,* vol. 21 (1980): 751-772.

Rindos, D. *The Origins of Agriculture, an Evolutionary Perspective.* Orlando, Florida: Academic Press, 1984.

Sauer, C. O. *Agricultural Origins and Dispersals.* New York: American Geographical Society, 1952.

Sax, F. M. "The Cytogenetics of Facultative Apomixis in *Malus* Species." *Journal of the Arnold Aboretum,* vol. 40 (1959): 289-297.

Solecki, R. S. "Prehistory in Shanidar Valley, Northern Iraq." *Science,* vol. 139 (1963): 179-193.

Stebbins, G. L., Jr. "Apomixis in the Angiosperms." *Botanical Review,* vol. 7 (1941): 502-542.

Stebbins, G. L., Jr. *Variation and Evolution in Plants.* New York: Columbia University Press, 1950.

Turner, B. L., II, and C. H. Milsicek. "Economic Plant Species Associated with Prehistoric Agriculture in the Maya Lowlands." *Economic Botany,* vol. 38 (1984): 179-193.

Ugent, D., S. Pozorski, and T. Pozorski. "Archaeological Potato Tuber Remains from the Casma Valley of Peru." *Economic Botany,* vol. 36 (1982): 182-192.

Vavilov, N. I. "The Origin, Variation, Immunity, and Breeding of Cultivated Plants." *Chronica Botanica,* vol. 13 (1949-1950): 1-364.

Whiting, A. F. "Ethnobotany of the Hopi." *Museum of Northern Arizona Bulletin,* vol. 15 (1939): 1-120.

Legacy and Stewardship

H. L. Sawatzky

Early in May of 1975, in a context not out of keeping with his *Northern Mists* (1968), Carl Sauer wrote to me in response to some findings I was then preparing for publication:

> . . . I enjoyed your solution of the arctic mirage, with vicarious pride in a lad who did something I couldn't have done. It was a grand job of seeing and solving . . . You'll not be having any lack of problems to interest you. A bump of curiosity is a rare inheritance; be thankful.

Coming, as it did, from a man whom I revered as friend and mentor, that was heady stuff indeed. Coming, as I did, from an upbringing within a stern, inflexible agrarian tradition of central European background, where curiosity and secular erudition were systematically discouraged by means of all the social and economic sanctions available to an essentially closed society, it emphasized to me that the often painful process of slipping the restraints imposed by that society had, after all, been most worthwhile.

"The land talks to you when you have an understanding of what it is that you are standing on and looking at," Sauer wrote (Hewes 1983, p. 144), in reference to physical geography, in 1960. I was to make his acquaintance as a newly arrived master's student the following autumn. He questioned me in great detail on all aspects of life in an "ethnic island" in the Canadian Prairies, but most particularly about the manner of utilization of the land.

I had grown up in the Agassiz Plain, on the subhumid eastern edge of the prairies, where a long winter and late seeding later placed maturing crops squarely in the characteristic July-August drought with the consequent inevitable progressive downward revision of yield prospects as the season advanced. The deep, fertile chernozems of the Agassiz Basin were regularly fallowed, to conserve moisture, it was said. Moldboard tillage, however, paradoxically, provided for practically no retention of moisture received as snow, while the characteristically high winds of late winter and spring played havoc with the pulverized, unprotected soil. My own grandfather had related to me how, as early as the 1880s, less than ten years after plow was set to sod, it was already common to "lose the planted seed out of the ground" by wind erosion. As a teenager and full participant in the labor of the family farm, I formulated notions of what might be called responsible stewardship, and agitated for a more conservationist approach to the modest acreage at our disposal. My arguments did not unseat the habituated usage of generations.

I would have been a farmer, and a farmer only, had there been the opportunity and the wherewithal. Years at various laboring jobs galvanized the determination to escape that routine, but just barely supplied the enonomic basis for that escape. Eventually I earned the credentials that brought me to Berkeley. The academic career that grew out of the Berkeley years made it possible to pick up where I had left off as a teenager and return to the development of concepts of utilization of the land and its intimately associated adjunct resources in a manner I believe Mr. Sauer would have approved of. What follows represents some of the insights and results which I have achieved to the present time.

Alternative Approaches to Energy and Water Management in Western Canadian Agriculture

At the latitude of the Canadian Prairies, in an interior continental setting open to the north and less than 5 degrees below the limits of permafrost, we are highly aware that growing season or, stated another way, energy supply is a major limiting factor to the productive potentials of agriculture. The growth potentials of plants are heavily dependent on temperature. Western Canadian farmers are frequently concerned with slow germination and lack of progress of crops during cool spells in May and June, despite a relative abundance of moisture. At other times, lack of moisture is the limiting factor and the cause of concern. Less obvious is the fact that frost may persist at depth in seeded fields until well into June.

The logical question that an appreciation of these realities poses is, "What, if anything, can be done, in an ecologically and economically sound way, to ameliorate the prospects of moisture excess, moisture deficiency, and energy deficiency, with a single set of basic yet flexible resource management techniques?" The central issue is, "Can we enhance the potentials of our 'biological crop-growing season' with the resources at hand, in an economically viable and technically uncomplicated way?"

This presentation reflects investigative experience and insights gained at field level, on a private, commercial production scale, through the application of a set of resource use theories centering on snow management. Investigations to refine and amplify the data are continuing.

In the context of the Prairie Provinces it is worth noting that in all of agricultural western Canada east of the Rockies there is, with only a few localized exceptions, no persistent excess of precipitation over potential evapotranspiration. Off-season precipitation captured and stored where it falls will, therefore, almost everywhere, be a valuable asset during the subsequent growing season. Snow, as a major part of off-season precipitation, therefore should have substantial resource appeal arising from an appreciation of this fact alone. Indeed, it was from this limited perspective that the investigations of the resource potentials of snow were begun. Since that time, ancillary potentials have been recognized and, in broad terms, interpreted and exploited. Considerable study of the insulating properties of snow has been carried out in both America and Eurasia (e.g., Thompson 1934, Rikhter 1945, Bay, Wunnecke, and Hays 1952, Chekotillo 1955, Shimanovskii 1955, Willis et al. 1961, Grant and Ramirez 1975, Gauer 1980).

The current high level of interest in the production of fall-seeded wheat on the Canadian Prairies has spurred research on this subject. The potential of snow as a source of moisture for subsequent crop production has also been the subject of considerable study (Staple, Lehane, and Wenhardt 1960, Willis et al. 1961, Gray, Norn, and Dyck 1970, Frank and George 1975, Willis and Frank 1975,Steppuhn 1976, Granger et al. 1977). Moreover, snow proffers both passive and dynamic energy conservation potentials as an incidental benefit of the exploitation of its insulating properties. These potentials appear to have been generally overlooked heretofore, and are the basis of my research.

The practical appreciation of the insulating properties of snow in agriculture has hitherto centered primarily on its function in maintaining a root-zone temperature—and in inhibiting temperature fluctuations—compatible with the survival of overwintering crops. This approach overlooks the simultaneously operative function—insofar as frost penetration is thereby reduced—of retention of the latent heat of crystallization in all that portion of the soil moisture which otherwise would have undergone the phase change to ice but, due to the insulating effects of snow cover, did not. Insofar as soil moisture does not undergo the phase change from liquid to crystal, the latent heat of crystallization (80 Kcal/kg) need not be replaced from the ambient energy supply at the onset of the following growing season. The differences involved can be quite dramatic.

As a case in point, two adjacent fields on Red River Clay loam monitored during the winter and spring of 1981-82 may serve to illustrate. One, 20 ha in extent, was in corn in 1981. It overwintered with an average standing stover height of 50 cm. In late March it had a settled snow cover of 30 cm over 15 cm of frozen ground. The other field, 24 ha in extent, was in peas in 1981, cut close to the ground and subsequently deep-tilled and harrowed down. Snow cover was discontinuous, but averaged approximately 2 cm, on ground frozen to a depth of just over 2 meters. The second field, therefore, had approximately 1.85 cubic meters more volume of frozen ground per square meter of surface. Total precipitation during September-November of 1981—almost all of it rain, or snow which melted prior to freeze up—was 149.6 mm, as measured by Environment Canada less than 5 km from the field's location. It may be assumed, therefore, that both fields were at field capacity at the onset of winter.

At these values, frost penetration on the cornfield involved 150 liters of soil mass per square meter of surface, with a frozen moisture content of 56 kg, the latent heat value of which was (80 x 56 = 4,480) on the order of 4,500 Kcal. The latent heat value of 30 cm (300 liters per square meter) of settled snow cover, at a water content of .2, works out to (300 x .2 x 80) 4,800 Kcal. Total latent heat replacement requirements per square meter were thus on the order of (4,500 + 4,800) 9,300 Kcal to achieve full thawing of all frozen moisture. Frost penetration on the peafield, by comparison, involved some 2,000 liters of soil mass per square meter of surface, with a frozen moisture content of 750 kg, the latent heat value of which was (750 x 80) 60,000 Kcal. Latent heat replacement cost of melting 2 cm of settled snow with a water equivalent of .2 would be (20 liters x .2 x 80) 320 Kcal per square meter. The difference in energy cost drawn from the ambient energy supply to replace the latent heat of crystallization in the course of total frost dispersal in the soil profile was therefore (60,320 – 9,300) on the order of 50,000 Kcal per square meter. The average net daily radiation at this season is approximately 4,600 Kcal/day. The 1982 crop seeded into the cornfield therefore had the benefit, within the same time-frame, of more than ten days' net radiation relative to the peafield, in which the energy was diverted to restoring latent heat to frozen moisture in the ground without affecting the ambient temperature of the environment.

An argument frequently advanced relative to comparisons such as this is that the differing albedos of dark, bare ground and light-colored trash- and snow-covered ground might to a large extent offset, through differing energy absorption rates, the differences in energy requirements calculated for the replacement of latent heat. Certainly bare soil most nearly approximates black body characteristics overall. The difference is significant, however, only in relation to direct short-wavelength solar radiation. In relation to long-wavelength reflected or advected energy or that conducted with rain, all colorations have equivalent energy absorption properties.

Repeated tests over several seasons have revealed that retreat of the upper frost horizon on bare ground is not more rapid than under any level of standing cover up to 50 cm. Indeed, it tends to be marginally slower, despite the additional energy demand for melting snow captured by the ground cover. This is not to suggest that energy absorption occurs at higher rates on snow-covered ground than on bare soil. On the contrary, total energy absorption on the bare, dark-colored soil *must* exceed that on covered ground. However, once melted, moisture beomes potentially mobile, and is drawn toward the surface. The relatively unimpeded, and hence rapid, exchange of air at the bare surface ground-air interface predictably, and at much higher rates than on covered ground over which air exchange is substantially impeded, results in *evaporation,* at an energy cost, at ambient temperatures, in excess of 600 Kcal/kg as compared to 80 Kcal/kg for melting. As latent heat is replaced at the upper frost horizon, therefore, moisture undergoing the phase change from frozen to liquid will tend to be evaporated at the surface, with much of the available ambient energy thus diverted into heat of vaporization, at an energy cost, per unit of water mass, of nearly eight times that involved in replacing latent heat of crystallization in the course of frost dispersal. It is evident that in this process the soil is not warmed but refrigerated. Fallow or other bare ground, therefore, contrary to still widely held popular belief, performs poorly in relation to effective energy absorption and moisture retention, and is at the same time ineffective as a land management concept in terms of energy flow characteristics.

Systematic treatment of snow as a resource tends to provoke several other resource effective—and, additionally, economically attractive—management adaptations. Erosion-suppression is, of course, an obvious and automatic benefit. Recognition of the crop response potentials inherent in the transportation of snow moisture into the growing season suggests the amplification of total field storage capacity by augmenting the humus content via subsequent soil incorporation of the snow-holding biomass. An ancillary benefit of this practice over time is to raise infiltration rates to a level more nearly equal to potential summer precipitation rates, thus systematically reducing ponding and drownout tendencies during the growing season as well as reducing drainage requirements and total annual runoff. Maximization of snow retention as a management goal invites critical evaluation of other management aspects. Straight combining, with the purpose of leaving standing stubble height geared to snow retention rather than to swath mass and stubble rigidity to carry the swath, is highly time and energy effective, and is readily possible up to the upper limits of "tough" conditions, with reduced losses in respect both to mechanical handling and loss of quality. Moreover, straight-combined fields, because of the higher stubble height possible with this technique, retain good snow-capture potentials even when subjected to deep tillage, should this be deemed desirable for weed suppression. Forced-air drying, using, in many parts of western Canada, renewable hydroelectric energy, can substitute in large part for nonrenewable fossil fuels in both harvesting and conditioning of the crop. Effective snow management is highly compatible with a minimum- and/or zero-till approach to land management.

In the program initiated by myself in 1978, overall machine time involved in tillage has been reduced by some 50 percent, thus greatly reducing the impact of various related cost escalations which have occurred since then. The only aspect of the cropping picture in which little overall time and energy reduction has been registered has been in the handling of the harvest, but this has been a function of the attractive yields recorded. Field-scale tests, undertaken on a comparison basis as strict as the scale of operations permitted, suggest that the yield response that can be expected from an increment of snow moisture is on the order of two times the response from an equivalent increment supplied by supplemental irrigation in the context of a winter-bare

209

or minimum-cover tillage regime. Even at accepted values for irrigation supplements, under conventional tillage, of approximately 100 kg of wheat or its equivalent per ha for each available 10 mm above 140 mm (Hedlin 1980), 60 mm of recorded snow moisture could have accounted for 600 kg/ha of the noted yield difference of up to 1,550 kg/ha in spring wheat. The prospect suggests itself, therefore, that a major hitherto unrecognized variable is represented by over-winter loss, through frost penetration, of massive amounts of latent heat from soil moisture in the conventionally tilled situation, and its large-scale retention in the other. This supposition appears all the more defensible in light of the fact that energy—or the lack of it—is generally recognized as a seriously limiting factor in the production of desirable economic plants in the context of the climatic realities of this region, where the normal frostfree season is 125 days or less, and even in most-favored localities has been as short, recently, as 108 days (1982).

Field-scale, farm-level experimentation (currently being followed up by replicated plot tests on similar soils at the University of Manitoba's agricultural research facility at Glenlea, Manitoba) has, as indicated, yielded substantial perceived benefits in both the input and output aspects of crop production. The energy-saving cultural practices suggested can be implemented by any farm operator, in most cases with the equipment already on hand. The economic results are attractive. Significant reductions in field time (and hence operating expense) have been accompanied by very satisfactory yields which in all cases have been greatest where snow retention had been most uniformly massive. With snow retention on the order of 25-30 cm yields approaching 5,000 kg/ha in barley, exceeding 4,000 kg/ha in hard red spring wheat, 2,400 kg/ha in flax, and 3,300 kg/ha in peas have been achieved. In each case, comparison fields under a conventional tillage regime involving little—and only incidental and discontinuous—snow capture experienced yields lower by an amount sufficient to reinforce substantially the perception that, in consequence of the inherent implications relative to the energy budget, an increment of moisture derived from *in situ* captured snow has a plant response value on the order of two times that expected from a similar moisture increment supplied as supplementary irrigation.

The fact that almost all of the agricultural area of western Canada experiences precipitation totals that are well below the optimum for crop response suggests that too high a priority has been set upon disposal of occasional, temporary water surpluses—in the main those associated with the spring melt. That this has contributed substantially to major flooding events in the lower reaches of the Red-Assiniboine drainage basin and in other prairie drainages is by now generally recognized. Of the 284,000 sq. km comprising the Red-Assiniboine basin 52,000 lie in Manitoba. The *in situ* retention of snow sufficient to yield 25 mm of supplemental field-stored moisture recharge over this area would aggregate to over 1.2 billion cubic meters. This amounts to between 15 and 20 percent of the total average annual discharge from the entire basin. It is also enough to inundate the area of the Agassiz Basin flooded in 1979—930 sq. km—to a depth of over 1.4 m. Viewed in the compressed time context of major flooding events of recent times, a reduction of total flow, of the magnitude potentially within the power of Manitoba to achieve independently within the larger basin context, would amount to more than 25 percent of the water carried by the Greater Winnipeg Floodway during a major flood threat such as that of 1979. Converted into crop response, this amount of water as the uniformly distributed equivalent of supplementary irrigation should have the potential for generating production equivalent to approximately 250 kg of wheat/ha or, in the Red-Assiniboine basin context of Manitoba, a total of more than 1.1 million tonnes. Even at current depressed prices, the value of such a production increment—achieved, moreover, in the context of substantially reduced unit production costs—would be on the order of $150 million (in Canadian dollars). It would appear, therefore, that the real cost of a flood such as that experienced in 1979 involved not only $30 million or so in assessed, visible damage, but a further lost opportunity cost at least five times as great, in Manitoba alone. A 25 mm increment in retained snowmelt over the cultivated portions of the entire basin would very probably eliminate the historically chronic flood threat entirely, or at the very least draw it down to the level of occasional inconvenience.

Finally, it is a truism that, in the context of plant response, moisture and energy are interdependent variables, the relative value of each being dependent upon the availability of the other. Expanding the available amounts of both through systematic inclusion of snow in the overall management of the agricultural land resource suggests strong incentives for the withdrawal of fallow from existing crop rotations. Moreover, the yield increments attainable under the suggested management regime with its diminished production costs, could go far toward closing the gap between currently realized yields and those obtainable under irrigation with attendant greatly augmented costs and substantial environmental risk. Indeed, it appears conceivable that future production goals might be achieved and sustained on the existing, primarily higher class, agricultural land base, without the necessity of invading and, at great cost, ameliorating, but quite possibly also permanently mutilating, the marginal lands in Soil Classes IV (severe limitations for agriculture) and V (very severe limitations for agriculture), which represent not only much of the remaining wildlife habitat but, in great measure, the aesthetic quality of our rural landscape as well.

Literature Cited

Bay, C. E., G. W. Wunnecke, and O. E. Hays. "Frost Penetration into Soils as Influenced by Depth of Snow, Vegetation Cover, and Air Temperatures." *Transactions,* American Geophysical Union, vol. 33 (1952): 4.

Chekotillo, A. M. "Temperature Inversion of Soil During the Cold Period." *Problems of the Study of Snow and Its Use in the National Economy.* Moscow: Academia Nauk S.S.S.R., 1955.

Frank, A. B., and E. J. George. "Windbreaks for Snow Management in North Dakota." *Proceedings,* Snow Management on the Great Plains. Publication 73. Lincoln, Nebraska: U.S.D.A. Agricultural Experiment Station, 1975.

Gauer, L. E. "Soil Temperature and Moisture of Conventional- and Zero-Tilled Soils in Manitoba." M.Sc. Thesis, Department of Soil Science, University of Manitoba, 1980.

Granger, R. J., D. S. Chanasyk, D. H. Male, and D. I. Norum. "Thermal Regime of a Prairie Snowcover." *Soil Science Society of America Journal,* vol. 41 (1977): 839-842.

Grant, L., and J. M. Ramirez. "Climatology of Snow in the Great Plains." *Proceedings,* Snow Management on the Great Plains. Publication 73. Lincoln, Nebraska: U.S.D.A. Agricultural Experiment Station, 1975.

Gray, D. M., D. I. Norn, and G. E. Dyck. "Densities of Prairie Snowpacks." *Proceedings,* Western Snow Conference. Fort Collins: Colorado State University, 1970.

Hedlin, R. A. "The Place of Summerfallow in Agriculture on the Canadian Prairies." *Proceedings,* Prairie Production Symposium, 1980.

Hewes, L., "Carl Sauer: A Personal View." *Journal of Geography,* vol. 82 (1983): 140-147.

Rikhter, G. D. *The Snow Cover, Its Formation and Properties.* Moscow: Izdatel'stvo Academia Nauk S.S.S.R., 1945. (English translation by SIPRE, Corps of Engineers, U.S. Army, Translation No. 6).

Sauer, C. O. *Northern Mists.* Berkeley: University of California Press, 1968.

Shimanovskii, S. V. "Thermal Regime of the Snow Cover." *Problems of the Study of Snow and Its Use in the National Economy.* Moscow: Academia Nauk S.S.S.R., 1955.

Staple, W. J., J. J. Lehane, and A. Wenhardt. "Conservation of Soil Moisture from Fall and Winter Precipitation." *Canadian Journal of Soil Science,* vol. 40 (1960): 80-88.

Steppuhn, H. W. "Areal Water Equivalents for Prairie Snow Covers by Centralized Sampling." *Proceedings,* Western Snow Conference, 1976.

Thompson, W. A. "Soil Temperatures at Winnipeg, Manitoba." *Scientific Agriculture Canada,* vol. 15 (1934): 4.

Willis, W. O., C. N. Carlson, J. Alessi, and H. J. Haas. "Depth of Freezing and Spring Runoff Related to Fall Soil Moisture Level." *Canadian Journal of Soil Science,* vol. 41 (1961): 115-123.

Willis, W. O., and A. B. Frank. "Water Conservation by Snow Management in North Dakota," *Proceedings,* Snow Management on the Great Plains. Publication 73. Lincoln Nebraska: U.S.D.A. Agricultural Experiment Station, 1975.

PART SIX
A Summary

Carl Sauer:
Explorer of the Far Sides of Frontiers

Alvin W. Urquhart

> *We shall not cease from exploration*
> *And the end of all our exploring*
> *Will be to arrive where we started*
> *And know the place for the first time.*
> T. S. Eliot *Four Quartets*

Frontiers of settlement have both a far side and a near side. The far side lies beyond the leading edge of settlement and is often little known or of little concern to those moving into the area. The near side becomes familiar to the newcomers who, because they are more powerful, remake and redefine the area in their own terms. Carl Sauer stood apart in his persistent concern for the people and places found on the far side of the European settlement frontiers in America. By comparing near side with far side, Sauer saw clearly the changes that follow in the wake of the passage of a frontier. He showed us the real changes wrought on lives and lands that were overwhelmed; he did not simply repeat the self-justifying assumptions of progress, advancement, or rightness expressed by those in control. His sympathies lay not with the powerful on the frontiers' near sides but with the soon-to-be engulfed countrysides and country folk on the far sides of the frontiers.

A "son of the Middle Border," Sauer grew up seeing the rag-tag ends of the European agricultural frontier which completed its sweep across North America at the end of the nineteenth century. Indian lands and cultures had largely disappeared as their far-side homelands had been converted to near-side America. This was the context in which Sauer began his first geographic research in Middle Western geography (1916, 1920, 1927). At the time, he was principally concerned with contemporary American landscapes, but he had also studied the advance of the frontier of American agricultural settlement and even had given a few nods toward future development. These studies formed the basis in the field for "The Morphology of Landscape" (1925).

Frontiers of Settlement

Sauer had to move away to gain a whole new perspective on landscapes and how they change. His strong interest in frontiers began to evolve only after he had moved to Berkeley and had completed fieldwork in the American Southwest. There he discovered evidence of a remote sixteenth- and seventeenth-century frontier between Spaniards and Indians. He wrote:

> The move to California in 1923 opened larger horizons of place and time. Field studies began in Lower California and continued in Sonora and Arizona. . . . The northwest of New Spain had been a Jesuit mission province comparable to Paraguay, and of similar age, from the late sixteenth to the Jesuit expulsion in the eighteenth century. . . . From arid north Mexico we went on to the tropical lands of the Pacific coast, using Spanish accounts of their condition in the past. By chance we came upon a forgotten area of high Mesoamerican culture well preceding the Spanish conquest (1980, p. 10).

Sauer retraced (in reverse direction) the old Spanish road to Cíbola, eventually ending up in Colima. All along the way he was reconstructing, in his mind, the native conditions before Cortés and his successors had expanded the Spanish frontier northwestward (West 1979). From Colima he climbed the Pacific slope to the Mexican highlands (1941). He wrote:

> [Then] it seemed proper to turn to the first Spanish entry into the New World by similar inquiry. What was seen and thought of the new lands and their inhabitability, how control was taken, and how possession was extended. Important lineaments of Spanish America were well drawn before Cortés began his march into Mexico [or Pizarro into Peru]. That span of a scant thirty years of discovery and domination gave not only geographical but large economic and political directions to Spanish Empire (1966, p. v).

In further attempts to understand the far side of the Spanish frontier before it reached Mexico, he directed his research to the Caribbean lands (1962b):

> I made use of what I had seen during a field season in the Dominican Republic, in travels about Cuba, Jamaica, Puerto Rico, and the Lesser Antilles, and in visits to the mainland coast. With this background of observation in mind I then studied the early reports as to places, plants, animals, and people, items peripheral to the interest of historians of the Spanish Indies. It was possible to trace the routes taken by the Spanish and to identify what they found, beginning with the Greater Antilles of Arawak population and continuing to the Isthmus of Panama, Chibchan in culture. The records gave extensive information on the various native economies and societies. Starting with Columbus, prosperous and well-balanced native ways of life were broken down in a few years, and a population of millions reduced to the point of extinction, all of this competently documented by Spanish sources (1980, pp. 10-11).

The Early Spanish Main (1966) gives the results of this study in a powerful, far-sided view that is quite different from those presented by most interpreters who have viewed the frontier largely from the Spanish (or near) side. After satisfying himself as to the nature of the far side of the Spanish frontier in America at the beginning of the sixteenth century, Sauer turned his attentions to the perceptions of the far side of the European frontier that had drawn Columbus and some of his predecessors westward.

The Europeans on the near side may have thought they would see Antilia, an island shown on many fifteenth-century maps. Antilia was but the latest of a series of legendary islands imagined by peoples of the western shore of Europe who looked seaward for fishing, commerce, adventure, or a new life. The real far side of the Atlantic frontier of Europe, however, was the land that was soon to be called the New World. And it was reports of these lands that attracted Sauer. In *Northern Mists* (1968b), he explored the seas and shores of the northern and western Atlantic, finding that in the fifteenth century the Portuguese had sailed northwest from the Azores to North American lands, but they in turn had probably been preceded by "men of Bristol, familiar with the Azores, Madeira and mainland Portugal" (1968b, p. 195). Even earlier, ships of the Hanseatic League, in the north, and of the Hermandad de las Marismas, from the Bay of Biscay, were fishing and whaling the North Atlantic. And before them, in the Atlantic frontier, were the Norse, who according to Sauer, "followed, mainly or entirely, routes of Irish discovery" (1968b, p. 195). The Irish in the tenth century:

> . . . went out widely as missionaries, pilgrims and hermits to be first to settle Faeroes and Iceland and, by slender but consistent evidence, to be the first Europeans to reach and remain on the North American mainland. What lay beyond the known horizon attracted them as pious romantics (1968b, p. 194).

How like Sauer to search for what drew the Spanish to the far side of the Atlantic frontier and, with no break at all, end up in North America in the tenth century with romantic Irishmen! And because the question, "What was on the far side of the frontier?" still called him, Sauer returned to explorers' accounts of the sixteenth century in search of answers. Although in earlier studies he had asked questions about pre-European North America (1927, 1950b, 1952), he had not yet tried to reconstruct more broadly the North American Indian land and life at the time just preceding the establishment of a lasting settlement frontier between Europeans and Indians in North America. (His school text, *Man in Nature: America before the Days of the White Man: A First Book in Geography* (1939), merely hints at this theme.) After he retired from teaching, he searched the many reports filed in archives and the travel accounts of North America just as he had done earlier for Mexico, the Spanish Main, and the fringes of the North Atlantic. In *Sixteenth Century North America* he wrote:

Land and life as observed by these early reports provides various and mutually supporting data to reconstruct the conditions before Europeans came . . . The composite of observations is sufficient to outline the geography as it was when Europeans came. Mainly it appeared a land favored by nature and well inhabited by natives, differing in their usages but of good habits and presence, and hospitable (1971, pp. x-xi).

Sauer's interpretations of the sixteenth-century accounts of Spanish, French, and English explorers were later followed by his picture of the far side of the French and Spanish frontiers in seventeenth-century North America.

With the study of seventeenth-century North America, Sauer saw that he had returned to the scene of his first geographic studies:

The view presented here of seventeenth century North America as it was known to the French and Spanish has drawn me back to the experience of my early years, going back to 1910, when I was beginning to learn geographical observations in the Illinois Valley. . . . I have known the Mississippi Valley, the Great Lakes, and the Southwest widely and, I think, well enough to recognize the geography as it was before the changes brought by civilization (1980, pp. 10-11).

Frontiers of Technologic Change

In a completely different sense of the term ''frontier''—as in frontier of knowledge—Sauer consistently tried to understand the conditions of land and life that preceded basic, early technologic changes such as the development of agriculture, horticulture, domesticated plants, specialized methods of fishing, hunting, and collecting, the use of fire, and even the humanization of the anthropoid animal. Sauer's contribution to knowledge was his search for the far side of frontiers of change; the time or period of change is here likened to the edge of a frontier. He found earlier habitats and landscapes on the far side; on the near side the technologically altered landscape existed. Just as his odyssey in following the far side of the frontiers of settlement started in the American Southwest, so did his search for the far side of technologic change. He wrote of his start:

Indians of Sonora, Papagos, Opatas, and Mayos gave me the first lessons not only in what they grew, the common threesome of corn, beans, and squash, but also in varieties particular to those parts and retained from the prehistoric past, as were their ways of storage. Thus began a learning that cultivated plants are living artifacts of times past, available where archeology and written documents are wanting, or making these more explicit.

Observations, including those of Spaniards who had come early, and a little archeology, gradually fitted into a geographic pattern of the New World that appeared to show two major regions of plant and animal domestication. In the northern one crops were grown for their harvest of seeds by the planting and selection of seeds. From the Caribbean south, the dominant procedure was vegetative reproduction . . . (1972, p. vii-viii).

In a series of articles starting with "American Agricultural Origins" (1936), Sauer began to search for ever earlier indications of agriculture, first in the Americas and then in the Old World, always fascinated by the changing relationship of humans to the organic world around them. That these early forms of cultivation had emanated from hearths or homelands remote in both time and place emerges in Sauer's works when he, as was his wont, took a far-sided rather than a near-sided perspective. His reconstruction of the far side of both time and place suggests that the earliest home of domesticated plants was the tropical lands of southeast Asia rather than the alluvial valleys of southwest Asia, the obvious choice when viewing the evidence only from the near side where much of what-had-gone-before was lost in both natural and technological changes (1936, 1947, 1950a, 1952, 1958, 1959, 1960). Not content with merely 10-20,000 years of agricultural time, Sauer consistently pushed backward through the Pleistocene, at each stage trying to recreate habitats of the far side of the frontiers of change that would have made possible the development of the innovations that eventually allowed humans to occupy all of the earth (1944, 1948, 1956a, 1961, 1962a, 1964, 1968a, 1970, 1975).

Two Paths Beyond the Frontier

I have tried to show that Carl Sauer's odyssey took two paths in search of what lay beyond the near side of the frontier—the side that was closest to home. The first path eventually led him back to his early field experiences in the American Middle West. The second path led him back to the first human habitats, the seashores of East Africa at the beginning of the Pleistocene. During his lifelong journey along these two paths, Sauer was always aware of the near side, both of American settlement and of technological change. This awareness complemented and enriched his views from the far sides of the frontiers. His studies of the "Settlement of the Humid East" (1942) and its longer version, "European Backgrounds of American Agricultural Settlement" (1976) which in turn is a detailed expansion of his 1929 study, "Historical Geography and the Western Frontier" (1930), record views of the near side of the American settlement frontier. The final leg of Sauer's path home is reported in "Status and Change in the Rural Midwest—A Retrospect" (1963a), which was written a few years after he had written of the near side of the Middle Border as it had been in the nineteenth century (1963b).

What a grand sweep through geographic space! A journey outward from the Ozarks of Missouri and the Middle West to the American Southwest, Mexico, and the Spanish Main, then into the North Atlantic and finally a return, through the eyes of early European explorers, to North America and the European frontier as it moved through the American East and into the Middle Border. And what a grand sweep through time! From contemporary America, back through the beginnings of agriculture and the use of fire, to the very beginnings of humankind.

Frontiers Compared

A view of the far side of frontiers gave Sauer, more than most other scholars, a better basis for knowing the changes effected on the near side of frontiers. His comparisons destroy any smugness that we might have felt about the progress of civilization. His views of the destructive exploitation caused by modern colonial expansion (1938a) and the ideas he developed in "Theme of Plant and Animal Destruction in Economic History" (1938b) seem as if they could have been written today and still be in the forefront of environmental thought. His "Agency of Man on Earth" (1956b) is, after thirty years, as good a chronicle of the near side of the world's frontiers of development when viewed with a far-sided perspective as is available. Sauer wrote:

> We remain a part of the organic world, and as we intervene more and more decisively to change the balance and nature of life, we have also more need to know, by retrospective study, the responsibility and hazards of our present and our prospects as lords of creation (1952, p. 104).

Sauer showed us the way and the value of retrospectively exploring frontiers. And now, more than ever, we need to explore the far sides of frontiers to give perspective to today's predominantly near-sided views.

Literature Cited

Sauer, C. O. *Geography of the Upper Illinois Valley and History of Development.* Bulletin no. 27. Urbana: Illinois Geological Survey, 1916.

Sauer, C. O. *The Geography of the Ozark Highland of Missouri.* The Geographic Society of Chicago, Bulletin no. 7. Chicago: University of Chicago Press, 1920.

Sauer, C. O. "Morphology of Landscape." *University of California Publications in Geography*, vol. 2, no. 2 (1925): 19-53.

Sauer, C. O. *Geography of the Pennyroyal.* Ser. 6, vol. 25. Frankfort: Kentucky Geological Survey, 1927.

Sauer, C. O. "Historical Geography and the Western Frontier." In *The Trans-Mississippi West; Papers Presented at a Conference Held at the University of Colorado, June 18-June 21, 1929,* edited by J. F. Willard and C. B. Goodykoontz, pp. 267-289. Boulder, Colorado: University of Colorado Press, 1930.

Sauer, C. O. "American Agricultural Origins: A Consideration of Nature and Culture." In *Essays in Anthropology Presented to A. L. Kroeber in Celebration of His Sixtieth Birthday, June 11, 1936,* p. 278-297. Berkeley: University of California Press, 1936.

Sauer, C. O. "Destructive Exploitation in Modern Colonial Expansion." *Comptes Rendus du Congrès International de Géographie, Amsterdam, 1938,* vol. 2 (1938a): 494-499.

Sauer, C. O. "Theme of Plant and Animal Destruction in Economic History." *Journal of Farm Economics,* vol. 20 (1938b): 233-243.

Sauer, C. O. *Man in Nature: America Before the Days of the White Man.* New York: Charles Scribner's Sons, 1939.

Sauer, C. O. "The Personality of Mexico." *Geographical Review,* vol. 31 (1941): 353-364.

Sauer, C. O. "Settlement of the Humid East." In *Climate and Man, Yearbook of Agriculture, 1941,* pp. 157-166. Washington: D.C.: Government Printing Office, 1942.

Sauer, C. O. "A Geographic Sketch of Early Man in America." *Geographical Review,* vol. 34 (1944): 529-573.

Sauer, C. O. "Early Relations of Man to Plants." *Geographical Review,* vol. 37(1947): 1-25.

Sauer, C. O. "Environment and Culture During the Last Deglaciation." *Proceedings, American Philosophical Society,* vol. 92 (1948): 56-77.

Sauer, C. O. "Cultivated Plants of South and Central America." In *Handbook of South American Indians,* edited by J. H. Steward, pp. 487-543. Vol. 6 (Smithsonian Institution, Bureau of American Ethnology, Bulletin 143). Washington, D.C.: U.S. Government Printing Office, 1950a.

Sauer, C. O. "Grassland Climax, Fire, and Man." *Journal of Range Management,* vol. 3 (1950b): 16-21.

Sauer, C. O. *Agricultural Origins and Dispersals.* Bowman Memorial Lectures, Series 2. New York: American Geographical Society, 1952.

Sauer, C. O. "Time and Place in Ancient America." *Landscape,* vol. 6 (1956a): 8-13.

Sauer, C. O. "Agency of Man on Earth." In *Man's Role in Changing the Face of the Earth,* edited by W. L. Thomas, Jr., pp. 49-69. Chicago: University of Chicago Press, 1956b.

Sauer, C. O. "Man in the Ecology of Tropical America." *Proceedings of the Ninth Pacific Science Congress, 1957,* vol. 20 (1958): 104-110.

Sauer, C. O. "Age and Area of American Cultivated Plants." *Actas del XXXIII Congreso International de Americanistas, San Jose, Costa Rica, 1958,* vol. 1 (1959): 213-229.

Sauer, C. O. "Maize into Europe." *Akten des 34. Internationalen Amerikanisten-Kongresses, Wien, 1960,* (1960): 777-788.

Sauer, C. O. "Sedentary and Mobile Bents in Early Man." In *Social Life of Early Man,* edited by S. L. Washburn, (Viking Fund Publications in Anthropology, no. 31), pp. 258-266. Tucson: University of Arizona Press, 1961.

Sauer, C. O. "Seashore—Primitive Home of Man?" *Proceedings, American Philosophical Society,* vol. 106 (1962a): 41-47.

Sauer, C. O. "Terra firma: Orbis novus." In *Hermann von Wissmann-Festschrift*, pp. 258-270. Tubingen: Geographisches Institut der Universität, 1962b.

Sauer, C. O. "Status and Change in the Rural Midwest—A Retrospect." *Mitteilungen der Oesterreichischen Geographischen Gesellschaft*, vol. 105 (1963a): 357-365.

Sauer, C. O. "Homestead and Community on the Middle Border." In *Land Use Policy in the United States*, edited by H. W. Ottoson, pp. 65-85. Lincoln: University of Nebraska Press, 1963b.

Sauer, C. O. "Concerning Primeval Habitat and Habit." In *Festschrift für Ad. E. Jensen*, pp. 513-524. Munich: Klaus Renner Verlag, 1964.

Sauer, C. O. *The Early Spanish Main*. Berkeley and Los Angeles: University of California Press, 1966.

Sauer, C. O. "Human Ecology and Population." In *Population and Economics*, edited by P. Duprez, pp. 207-214. Proceedings of Section V of the Fourth Congress of the International Economic History Association. Winnipeg: University of Manitoba Press, 1968a.

Sauer, C. O. *Northern Mists*, Berkeley and Los Angeles: University of California Press, 1968b.

Sauer, C. O. "Plants, Animals and Man." In *Man and His Habitat*, edited by R. E. Buchanan, E. Jones, and D. McCourt, pp. 34-61. London: Routledge and Kegan Paul, 1970.

Sauer, C. O. *Sixteenth Century North America: The Land and People as Seen by Europeans*. Berkeley and Los Angeles: University of California Press, 1971.

Sauer, C. O. *Seeds, Spades, Hearths, and Herds*. Third edition of *Agricultural Origins and Dispersals*, with supplement. Boston: MIT Press, 1972.

Sauer, C. O. "Man's Dominance by Use of Fire." *Geoscience and Man*, vol. 19 (1975): 1-13.

Sauer, C. O. "European Backgrounds of American Agricultural Settlement." *Historical Geography Newsletter*, vol. 6, No. 1 (1976): 35-58.

Sauer, C. O. *Seventeenth Century North America: Spanish and French Accounts*. Berkeley: Turtle Island Foundation, 1980.

West, R. C. *Carl Sauer's Fieldwork in Latin America*. Dellplain Latin American Studies 3. Ann Arbor: University Microfilms International, 1979.

Contributors

- **Homer Aschmann** is a Professor of Geography at the University of California, Riverside, where he has taught since 1954. His interests include cultural geography and the human impact on the natural landscape.

- **Henry J. Bruman** has been a Professor of Geography at U.C.L.A. since 1945. His interests include historical geography of Latin America, colonization and settlement, plant geography, and Alexander von Humboldt.

- **William M. Denevan** (Ph.D. Berkeley, 1963) is Professor of Geography at the University of Wisconsin, Madison. His research has focused on Indian and peasant agriculture in the Amazon and Andes.

- **Carl L. Johannessen** (University of Oregon) is a Latin Americanist geographer specializing in the domestication of plants and animals and the distribution of wild vegetation as modified by humans.

- **Martin S. Kenzer** is an Assistant Professor of Geography at Louisiana State University. His primary research interests are historical geography and geographic thought.

- **Anne Macpherson** received her Ph.D. from the University of California, Berkeley. In 1976 she was awarded an N.S.F. grant to organize the Sauer Papers.

- **Geoffrey J. Martin** (Professor of Geography, Southern Connecticut State University) is author of books on Mark Jefferson, Elsworth Huntington, Isaiah Bowman, and co-author with Preston E. James of *The Association of American Geographers: The First Seventy-Five Years* (1981) and *All Possible Worlds* (1981).

- **Kent Mathewson**, a Ph.D. candidate, University of Wisconsin, Madison (Geography), is currently visiting lecturer in Geography, University of North Carolina-Chapel Hill. He has published articles on prehistoric agriculture, and is the author of *Irrigation Horticulture in Highland Guatemala* (1984).

- **Marvin W. Mikesell** (Ph.D. University of California, Berkeley, 1959) is a professor of Geography at the University of Chicago.

- **James J. Parsons** received his undergraduate and graduate degrees from the University of California, Berkeley, and has been a member of its Geography faculty for his entire academic career.

- **H. Leonard Sawatzky** (Ph.D. Berkeley) is an Associate Professor of Geography at the University of Manitoba. His interests are agricultural colonization and resource management.

- **William W. Speth**'s interest in the nature and development of geographic ideas was stimulated by study under Clarence Glacken, John Leighly, and Carl Sauer at Berkeley.

- **Alvin W. Urquhart** (University of Oregon) who received his A.B., M.A., and Ph.D. at the University of California, was enticed into geography in 1952 by Carl Sauer's course, the Conservation of Natural Resources.

Index